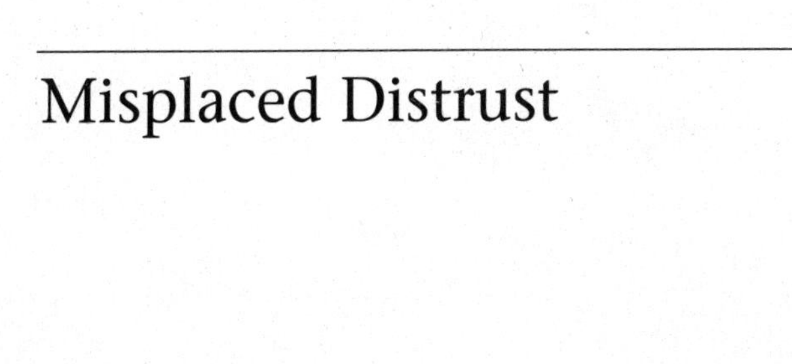

Misplaced Distrust

Éric Montpetit

Misplaced Distrust: Policy Networks and the Environment in France, the United States, and Canada

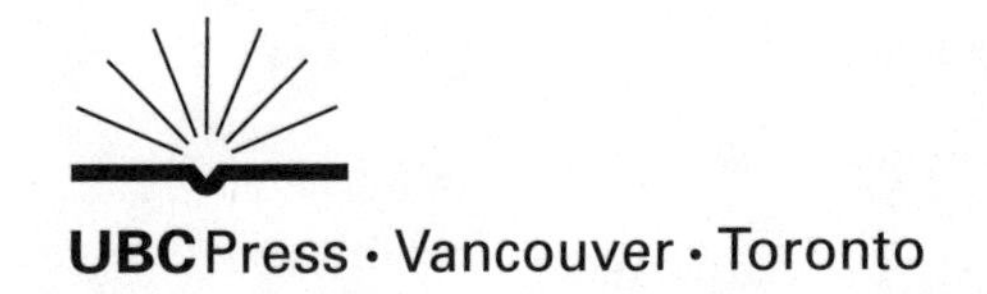

UBC Press · Vancouver · Toronto

09 08 07 06 05 04 03 5 4 3 2 1

Printed in Canada on acid-free paper ∞

National Library of Canada Cataloguing in Publication Data

Montpetit, Éric, 1970-
Misplaced distrust : policy networks and the environment in France, the United States, and Canada / Éric Montpetit.

Includes bibliographical references and index.
ISBN 0-7748-0908-6

1. Policy networks. 2. Environmental policy – Canada. 3. Environmental policy – France. 4. Environmental policy – United States. I. Title.
H97.7.M66 2003 352.3'7 C2003-910693-4

Canadä

UBC Press gratefully acknowledges the financial support for our publishing program of the Government of Canada through the Book Publishing Industry Development Program (BPIDP), and of the Canada Council for the Arts, and the British Columbia Arts Council.

This book has been published with the help of a grant from the Canadian Federation for the Humanities and Social Sciences, using funds provided by the Social Sciences and Humanities Research Council of Canada.

UBC Press
The University of British Columbia
2029 West Mall
Vancouver, BC V6T 1Z2
604-822-5959 / Fax: 604-822-6083
E-mail: info@ubcpress.ca
www.ubcpress.ca

À Geneviève

Contents

Tables / viii

Acknowledgments / ix

Abbreviations / xi

1 Introduction / 3

2 Assessing Policy-Making Performance / 20

3 Networks and Performance / 37

4 France: A Shift from Low- to High-Level Performance / 54

5 The United States: Performance in the Absence of Intergovernmental Coordination / 71

6 Canada: Stalled at a Low Performance Level / 90

7 Misplaced Distrust / 105

Notes / 123

Bibliography / 139

Index / 147

Tables

2.1 Environmental policy instrument types: Definitions, characteristics, and degree of epistemic community approval / 24

2.2 Level of agreement across three epistemic communities on the characteristics of agro-environmental policy / 34

3.1 Policy networks / 44

3.2 Internationalized policy environments and policy networks / 50

4.1 Agro-environmental policy instruments in France in the 1980s / 57

4.2 New or amended agro-environmental policy instruments in France in the 1990s / 60

5.1 Federal policy instruments in the United States and some aspects of state policies / 74

5.2 Changes in US federal and state agro-environmental approaches / 77

5.3 Livestock regulations in three US states / 81

6.1 Canada's federal and Ontario agro-environmental policy instruments / 92

6.2 Canadian agro-environmental policy approaches and the comprehensiveness/intrusiveness of their instruments / 96

7.1 Variations in constellations and networks / 114

Acknowledgments

I began thinking about distrustful attitudes toward the state early in the 1990s. At the time, in most countries belonging to the Organization for Economic Cooperation and Development (OECD), government deficits and debt were running so high that citizens legitimately feared policy makers had lost control. Economic growth was slow to pick up and unemployment was stagnating at levels rarely attained since the Second World War. Cynicism was understandable. Shortly thereafter, economic growth rates reached record heights, unemployment was in some places alarmingly low, government budgets were balanced or nearly so, and the burden of the debt was rapidly diminishing. Nevertheless, cynicism toward politicians, public sector employees, and, more generally, the state continued to flourish. Alarmed by this state of affairs, I decided to turn my attention to the arguments upon which distrust rests and, in the end, to challenge negative public opinions toward the state.

Because the public is often trusted to be right in democracy, it is rare for a book to claim directly that public opinion is wrong. Social scientists have often preferred to study the causes and consequences of public opinion. Motivated by the belief that social scientists also have a role to inform and thereby shape public opinion, I thus chose to write this book.

The Social Sciences and Humanities Research Council of Canada, the Fonds pour la formation des chercheurs et l'aide à la recherche, and the Fulbright Program contributed financially to the research leading to this book. I conducted much of this research while hosted as a guest scholar at Duke University and at the Institut national de recherche agronomique. I worked on the manuscript while teaching at the University of Saskatchewan and at the École nationale d'administration publique. I am indebted to several friends and colleagues from these institutions with whom I had fruitful discussions on several issues covered in this book. William Coleman was a great source of encouragement and inspiration. The comments of Michael Atkinson, Kathryn Harrison, and Mark Sproule-Jones on an early draft offered solid guidance for the preparation of the final manuscript. Research assistance by Francis Garon, Nathalie Dubois, and Lina Francoeur was of great value. Tony Nuspl, Linda Burr, and Carole Garant provided me with crucial editing and formatting assistance in the final stages of manuscript production. I want to thank Peter Lipert and Lora Hutchison, who generously accepted an invitation to brainstorm on a title for the book on a Saturday evening. Last but not least, Geneviève Bouchard (my girlfriend when I began this book and now my wife) has been my greatest source of encouragement throughout this project.

Abbreviations

International

CAP	Common Agricultural Policy (Europe)
EAGGF	European Agricultural Guarantee and Guidance Fund
EC	European Community
EU	European Union
FAO	Food and Agriculture Organization of the United Nations
GATT	General Agreement on Tariffs and Trade
NAFTA	North American Free Trade Agreement
OECD	Organization for Economic Cooperation and Development

France

ANDA	Association nationale pour le développement agricole
CEDAPA	Centre d'étude pour un développement agricole plus autonome
CORPEN	Comité d'orientation pour la réduction de la pollution des eaux par les nitrates
CTE	Contrat territorial d'exploitation
FNSEA	Fédération nationale des syndicats d'exploitants agricoles
PDD	Plans de développement durable
PMPOA	Programme de maîtrise des pollutions d'origine agricole
ZES	Zones d'excédent structurel

United States

AFO	animal feeding operation
CAFO	concentrated animal feeding operation
CREP	Conservation Reserve Enhancement Program
CRP	Conservation Reserve Program
EPA	Environmental Protection Agency
EQIP	Environmental Quality Incentives Program
NPDES	National Pollutant Discharge Elimination System

NPPC	National Pork Producers Council
NRCS	Natural Resources Conservation Service
USDA	United States Department of Agriculture

Canada

AESI	Agricultural Environmental Stewardship Initiative
CARD	Canadian Adaptation and Rural Development
CEPA	Canadian Environmental Protection Act
CFFO	Christian Farmers Federation of Ontario
CURB	Clean Up Rural Beaches (Ontario)
EFP	Environmental Farm Plan (Ontario)
MEF	Ministère de l'Environnement et de la Faune (Quebec)
NDP	New Democratic Party
NFU	National Farmers Union
NPA	National Program of Action for the Protection of the Marine Environment from Land-Based Activities
OFA	Ontario Federation of Agriculture
OFEC	Ontario Farm Environmental Coalition
OMAFRA	Ontario Ministry of Agriculture, Food, and Rural Affairs
OMEE	Ontario Ministry of Environment and Energy
OSCIA	Ontario Soil and Crop Improvement Association
SWEEP	Soil and Water Environmental Enhancement Program (Ontario)

Misplaced Distrust

1
Introduction

Citizens in most countries of the world profoundly distrust policy makers and, more generally, state institutions. This attitude has taken root in developed countries even though their economic and social conditions are good, relative to the rest of the world. In an influential comparative study, Inglehart writes that "today, political leaders throughout the industrialized world are experiencing some of the lowest levels of support ever recorded."[1] Twenty-five years after the publication of *The Crisis of Democracy* by Crozier, Huntington, and Watanuki – a groundbreaking study predicting a bleak future for democracy in the Trilateral countries – Pharr and Putnam have assembled a number of authors to revisit the seminal thesis. While there is no consensus on the causes and the consequences of the declining confidence in policy makers, almost no one disputes its significance. In their introductory chapter to *Disaffected Democracies*, Putnam, Pharr, and Dalton write: "A large body of evidence demonstrates that over the quarter century since Crozier and his colleagues issued their report, citizen's confidence in governments, political parties, and political leaders has declined significantly in most of the Trilateral democracies."[2]

Studies have related this widespread distrust to several factors, including: increased media coverage of corruption scandals, unmet expectations and disappointment with the ideals of modernity, and changing social tastes. Highlighting bureaucratic aberrations and doubtful behaviour on the part of politicians now occupies a pride of place in the news media's marketing strategies.[3] This situation can serve only to foster cynicism among those citizens enduring unemployment and social inequities while they await the prosperity that modernity was expected to bring.[4] Furthermore, the younger generation born at the tail end of the age of modernity may not even share these materialistic values, which have served to legitimize contemporary state institutions.[5]

Whatever the specific reasons behind this crisis of confidence, it is broadly associated with a sense that countries are not properly governed.

Some people might believe corruption diverts policy makers from their legitimate objectives, while others might think state institutions are not designed to address the right problems, but all seem to agree that government is not what it should be. Putnam, Pharr, and Dalton put forward the following argument: "The fact that public confidence has declined can be taken to mean that governmental performance is less satisfactory than it once was."[6] Concurring with them, although approaching the problem from an entirely different theoretical perspective, Schneider and Ingram state:

> Discontent with democracy in the United States carries a curious twist in that criticism is not directed at the traditional symbols or mechanisms of democracy. Complaints are seldom heard about persons being denied the right to vote, to express their opinions, or to run for office. Journalists are not being jailed, threatened, or fined for criticizing the government or its policies. Leaders of radical social movements – either right or left – are not being denied the right to speak, demonstrate, or assemble. Criticisms in the United States center around *governance* – the capacity of a democracy to produce public policy that meets the expectations of the society.[7]

Concerns about governance are not only common in the United States, but widespread throughout the advanced world. These concerns lie at the root of my motivations for this book proposing a closer examination of governance, via a comparative perspective, in three advanced democracies: France, the United States, and Canada.

Studying governance means recognizing that policy making is not exclusively the purview of government since bureaucracies and interest groups are also involved. Citizens not only distrust government, but also distrust the bureaucracies and the various interest groups involved in policy making. As Pierre and Peters argue, "a key reason for the popularity of this concept [governance] is its capacity – unlike that of the narrower term 'government' – to cover a whole range of institutions and relationships involved in the process of governing."[8] In other words, governance is achieved through complex relationships within the policy networks of state and civil society actors. Policy networks are structures that regulate the interactions of state and civil society actors in the governance process. These networks can influence how problems are defined or how the agenda is set. Certainly, they play a key role in policy formulation and implementation.[9]

Political scientists have been prolific in developing theories to account for the role of state, nonstate, and even extraterritorial actors interacting in networks for the purpose of governing. Interestingly enough, several of these theoretical constructions inspire no more than further disillusionment with governance. Power-hungry bureaucrats and resourceful special interest groups, allied with extraterritorial organizations accountable to no one, are often presented as controlling an ever-increasing share of policy-making activities.[10] Interestingly, observers of these distrustful attitudes, although attentive, tend to overlook the effect that this type of intellectual production itself may have on the crisis of confidence.[11] In any case, this book seeks to assess whether, in line with such theories, networks of actors governing advanced societies fail to address so alarmingly the most serious collective problems. Are policy solutions so poorly designed as to worsen problems rather than make them better? In other words, are countries so badly governed as to justify distrust in policy-making institutions and actors? These questions will be examined through a careful study of agro-environmental policy, one field of policy making among many to which they could be addressed.

Producing serious empirical analysis investigating these questions is of crucial importance. Although this crisis of confidence has remained short of generating popular demonstrations and mass riots against democratic institutions, it is not without consequences. Distrust encourages illicit behaviour such as cheating on taxes and a more general disregard for the law. It also discourages young, talented people from contemplating a career in the public sector. Distrust reduces the possibility for social solidarity and may consequently cultivate social fragmentation. It gives momentum to the project of dismantling states' policy-making capacity. In so doing, distrust runs the risk of becoming a self-fulfilling prophecy since a harsh social climate and the hollowing-out of the public sector make for difficult conditions of governance. Given these plausible consequences, the scholarly output and the theoretical models of governance that feed into the crisis of confidence deserve serious analysis.

The detailed examination provided here of environmental policy development for the agricultural sector in three countries – France, the United States, and Canada – should serve as a warning against the temptation to rapidly conclude that governance is alarmingly inadequate in advanced democracies. More often than not, policy networks are found to produce agro-environmental policies that address the concerns of wide constituencies. What is more, poor policy making has little to do with big bureaucracy and special interest groups to which political scientists have often attributed policy failures. This analysis suggests that publicly engaged intellectuals should attempt to curb the confidence crisis before the self-fulfilling prophecy of inadequate governance becomes a reality.

Problem-Solving Policy Networks

Central to this book is the idea that a large proportion of the population in most countries believes governance does not occur in the way it should. Since, as mentioned above, governance in advanced economies is achieved through policy networks, the confidence crisis suggests that citizens must often view these networks as wholly inadequate. But what would an adequate policy network look like?

Fritz Scharpf argues that the legitimacy of democratic policy-making arrangements can be both input-oriented and output-oriented.[12] Input-oriented legitimization is based on the idea that democratic government is "government by the people." Based on the principle of popular sovereignty, it requires a citizenry that desires to participate in the making of policy choices and therefore calls for open political institutions. Allowing a wide participation, institutions must inspire a sense that each and every citizen counts in the political community; institutions must allow for the voice of any citizen to be heard.[13] In other words, an input-oriented perspective on legitimacy accords participation a value independent of the quality of the policy choices it produces.[14]

This is a highly idealistic perspective on legitimacy. Scharpf has argued that in large countries, where the citizenry is minimally unified, institutions can rarely simultaneously allow everyone to participate and still function adequately.[15] The idea of a policy network itself suggests a certain form of exclusion, which, it can be assumed, renders policy making easier. To be sure, openness can confer a measure of input-oriented legitimacy onto policy networks, but, as they imply some exclusion, participation in them will never constitute a sufficient source of legitimacy. Therefore, when examined exclusively from an input-oriented legitimacy perspective, networks never come across as adequate policy-making arrangements. In networks, as in most institutional settings, input-oriented legitimacy needs to be complemented by output-oriented legitimacy.

Where "government by the people" can be only approximated, citizens of democratic societies should expect "government for the people" to confer output-oriented legitimacy on policy-making arrangements. As Scharpf sees it, "output-oriented notions refer to substantive criteria of *buon governo,* in the sense that effective policies can claim legitimacy if they serve the common good."[16] In areas where specialized knowledge is required, all forms of popular participation can even be suspended and power delegated to administrative bodies that are thereby entirely deprived of input-oriented legitimacy. In several countries, central banks have very little input-oriented legitimacy, relying instead on output-oriented legitimacy. Citizens are content with central bankers looking after their long-term interests. The judicial review responsibilities of courts also fall into this category. Constitutional law is a fundamental area that citizens and elected officials in several countries have preferred to leave in the hands of those who possess the best knowledge of legal principles and traditions.

While central banks and courts sometimes make controversial decisions, the belief that worse decisions would come out of institutions more open to participation is a sufficient source of legitimacy.

Most of the time, however, policy efficacy does not require such a complete suspension of the institutions from which input-oriented legitimacy normally emerges. More often, input-oriented aspects blend with output-oriented aspects to confer legitimacy upon policy-making arrangements. In addition, it should not be assumed that policy efficacy necessarily decreases as participation increases. Scharpf suggests that mechanisms such as elections, which we tend to view in terms of input-oriented legitimacy, can sometimes contribute to increasing policy-making efficacy.[17] In a recent study, MacAvoy shows that ordinary citizens' participation in environmental management makes for better solutions.[18] And I later argue that good policy-making performances require a relative openness of policy networks.

Nevertheless, participation alone rarely appears sufficient to legitimize institutions. Results are always expected: While results sometimes justify the suspension of participation, participation never justifies the suspension of results. In other words, input can only go so far in legitimizing political institutions, and therefore output-oriented legitimacy appears crucial to modern states. Modern states need the capacity to resolve pressing collective problems. To return to the initial question in this section, an adequate policy network is one that makes a strong contribution to output-oriented legitimacy by resolving collective problems efficiently.

Surprisingly, few studies have sought to link policy network structures and the performances of political actors attempting to resolve problems. Most of the study of networks has preferred to shed light on participation, on patterns of exclusion, or on the power distribution embodied within networks. Rhodes suggests six reasons that make networks important: (1) they limit participation; (2) they define roles; (3) they exclude issues from the policy agenda; (4) they shape the behaviour of actors; (5) they privilege certain interests; and (6) they substitute private government for public accountability.[19] Rhodes' interest in policy outputs is limited; while conclusions on policy outputs may be derived from the power-interaction analysis he proposes, he does not go the extra mile to link networks to systematic policy choices.[20] In other words, the network analysis produced by Rhodes raises important input-oriented legitimacy questions but does little for output-oriented legitimacy.

The work of Bressers and O'Toole is more systematically concerned with the policy outputs that networks are likely to yield. Specifically, they attempt to show that the choice of policy instruments depends on two sets of network characteristics: (1) the degree of cohesion of the networks in question; and (2) the interconnection they establish between state and civil society actors. For instance, they argue that networks with weak cohesion and weak interconnectedness, because they require state actors to make an appeal to more coherent normative values, are likely to resort to regulations.[21] While this may represent one of the most systematic attempts to correlate network types and policy outputs, the focus on policy instruments does not extend easily to reflections on specific performances of political actors and policy networks or, as a result, to reflections on output-oriented legitimacy. As I argue in Chapter 2, it may be possible to attain the same result with two different policy instruments, at least in the environmental sector.

Moving closer to the question of output-oriented legitimacy is the network literature centred on policy change. At the heart of these studies is the idea that networks display different capacities for mediating the effect of globalization or regional integration. In countries and sectors where networks establish close interconnections between well-organized groups and state agencies, often in so-called corporatist networks, adaptation to regional or international pressure takes on a distinctive domestic character.[22] This conclusion encourages reflections on output-oriented legitimacy in many ways. Borrowing from Polanyi's *Great Transformation*, some might view as encouraging that resistance to the tenets of the self-regulating market remains possible in this era of globalization.[23] However, the proponents of the self-regulating market might have a different view. In any case, the objective of these studies on policy change has not been to think about output-oriented legitimacy systematically. In addition, globalization is only one among several challenges facing the modern state, and therefore assessing networks' adaptational capacity provides only a partial view of their performance.

When policy networks comprise strong state agencies, Atkinson and Coleman argue, they anticipate rather than react to problems.[24] Although their notions of anticipatory and reactionary industrial policy are not easily transferable to sectors other than economic policy, Atkinson and Coleman provide a relative if not all-encompassing view of the policy-making performance of networks. In line with this groundbreaking work, Weiss suggests that networks establishing a pattern of "governed interdependence" between the state and civil society are those most likely to be capable of developing successful policy strategies in an ever-changing world.[25] Although such networks constrain policy-making participation, they nonetheless constitute an important source of output-oriented legitimacy. Taking my cues from Weiss, I present in Chapter 3 specific hypotheses regarding networks' capacity to generate good policy-making performance. Establishing a clear relationship between networks and policy-making performance is an essential and distinctive feature of this book.

Assessing Policy-Making Performance

When policy networks function adequately, or when they are capable of resolving collective problems – in short, when their policy-making performance is good – they deserve some legitimacy. One might suggest citizens are being misled if they are told to distrust such governance mechanisms. But how does one go about assessing policy-making performance?

Policy makers address a broad range of problems, from unemployment to the depletion of fish stocks, each requiring very particular expertise and solutions. Given such conditions, the task of assessing the problem-solving performance of policy makers for a representative sample of problems would be impossible to accomplish in a single book. The more representative the sample of problems becomes, the fewer the details that can enter into the analysis. This is a common problem in social science; as the number of cases increases, the relevance of the analysis to each of them often decreases. In this book, I have therefore chosen to focus on a single policy problem, assuming that cases differ so much in nature that any addition would engender a significant reduction in the quality of the analysis. I have focused my attention on agricultural pollution as it presents itself in three countries: France, the United States, and Canada.

Unlike for some other social problems, a large body of knowledge exists on potential solutions to environmental problems. It is not my intention to minimize the difficulty associated with addressing environmental problems, but as a British official once said: "There is virtually no form of environmental pollution that we do not know how to control."[26] It is particularly important for such a study on policy makers' problem-solving capacity to focus on potentially resolvable problems. Who could blame policy makers for failing to solve problems for which no solutions currently exist? Agricultural pollution is indisputably a down-to-earth problem for which, as we will see in Chapter 2, several innovative solutions circulate in policy-making networks.

Inglehart's studies of changing social tastes also justify the selection of such a postmaterialistic problem as environmental protection.[27] As citizens' worries become increasingly distant from material concerns, policy makers should pay increasing attention to postmaterialistic problems. In fact, policy makers may not be deserving of trust if they limit themselves to addressing problems in which citizens are losing interest.

If agricultural pollution is a problem for which innovative solutions exist, adopting these solutions nevertheless involves facing up to serious difficulties. Policy networks, upon which the responsibility is likely to fall for developing effective agro-environmental policies, are unlikely to be well disposed toward innovative solutions. Agriculture is a sector in which state intervention has been high, historically speaking. In turn, for administrative purposes, state intervention has involved establishing strong agricultural bureaucracies accustomed to aiding farmers. Agricultural bureaucracies should consequently have a greater capacity than environmental agencies to develop and diffuse their policy preferences. In addition, farmers are generally efficiently organized and can therefore be expected to exert great influence within networks.[28] Notably, they should be capable of resisting environmentalists' demands for constraints on farming practices. Furthermore, those who no longer view policy making in exclusively domestic terms are also well served, as agriculture is arguably the object of the most ambitious common policy in Europe – a policy, incidentally, presently confronting certain difficulties[29] – and now constitutes a major stake in international trade negotiations. It is at best unclear whether this supranational policy-making context enables superior agro-environmental policy-making performance. These themes will be examined in greater detail in Chapter 3. Suffice it to say, for the moment, that if policy makers can successfully develop environmental policies for agriculture, there is no reason to expect them to be unable to attain similar results in the relatively numerous sectors where political obstacles are fewer. While it may be hazardous to generalize from such a specific case as agricultural pollution, revealing successful environmental policy development in agriculture would offer a solid indication that policy networks can resolve problems more generally.

Successful Environmental Policy Development in Agriculture

In this book, the successful development of environmental policy in agriculture serves as an indication that policy networks are capable of addressing collective problems and that, therefore, they are deserving of trust. But the notion of successful environmental policy development deserves clarification. Janice Gross Stein reminds us that success in the utilitarian tradition is a function of internal satisfaction.[30] If citizens who use the environment are satisfied, then environmental policy is successful. Inversely, if citizens who use the environment are unsatisfied, then the policy fails. In short, this utilitarian view forbids any questioning of citizens' views. Citizens just cannot be wrong: If they think governance is inadequate, it must be inadequate. As I see questioning the opinions provided by citizens surveyed about governance as a worthy enterprise, I naturally reject the utilitarian method.

But I am not alone in thinking this way. In an analysis of education policy, Stein argues that parent satisfaction cannot be the sole measure of policy performance. She writes: "Most [citizens] would expect schools to be held accountable for more than the satisfaction of parents. They would expect student achievement to improve over time."[31] In the introduction to a collective book that takes dissatisfaction with government very seriously, Putnam, Pharr, and Dalton write: "None of us argue that popularity is the sole measure of democratic performance, and all of us recognize that governments often must (or should) take actions that might reduce their popularity in the short run. Some of us believe that democracy is not (just) about making citizens happy, and that it is also supposed to facilitate 'good government,' whether or not citizens are pleased with government action."[32]

Just as consumers lack knowledge about products, citizens are rarely in a situation of perfect information about policy. Their opinions are often shaped by widely diffused impressions that rarely accurately reflect the actual situation in a sector. And it should also be underlined that public opinion itself is not a homogeneous whole. While I agree that the individual citizen's views should not be taken lightly (otherwise, I would not have written this book), they do not always provide a reliable and clear measure of policy success. Therefore, assessing policy-making success requires reference to collective values or external criteria. And reduction in the environmental impact of modern farming practices immediately comes to mind as the ultimate criteria for policy success in the agro-environmental sector. Unfortunately, criteria selection is rarely this simple, essentially because of measurement problems.[33]

First, deciding upon when a policy problem has been effectively resolved often depends on where one draws the boundary that defines the problem. For example, assuming that environmental experts have determined that the water in a hypothetical watershed shows unacceptable levels of contamination, one might suggest that the problem is solved the day the contamination is brought down to an acceptable level. Moreover, where environmental policies contribute to contamination reductions of this sort, analysts might be tempted to conclude that successful policy-making efforts have been made. However, a broader examination of the situation might reveal that contamination levels were not reduced without causing other problems. The environmental regulatory standards may have impaired the competitiveness of the watershed and caused job losses. Industries forced to invest in expensive technologies may suffer from a reduced capacity to employ people. In addition, reductions in water contamination in the watershed might cause new environmental problems if the technologies used to comply with water regulations shift pollutant emissions to the air and soil. In short, the broader view one has of a problem, the less likely one is to conclude that the solutions applied have addressed it in a satisfactory manner.

Second, it is a difficult task to link a "solved" problem – if one can convincingly be identified – to the effects of public policy. To illustrate this difficulty, Putnam speaks of the "Massachusetts Miracle Fallacy."[34] As he explains, despite politicians' rhetoric to the contrary, the strong economic performance in New England at the end of the 1980s was not caused by state policy. Similarly, reductions in contamination levels in the hypothetical watershed discussed above may be attributable to a wide range of factors, and public policy may be only one of them. The adoption of cost-efficient new technologies by industry can effectively reduce pollution in the absence of any state intervention.[35] Lower contamination may also result from plant closures due to a downturn in the business cycle that may not have much to do with government policy. Conversely, a situation where pollution levels remain high despite policy makers' best efforts may not necessarily indicate a policy "failure." When state intervention cannot improve a situation, it may nevertheless prevent it from getting worse and even contribute to avoiding major catastrophes. For example, public policies might incite growing industries to contain increases in pollutant discharges, thereby preventing a pollution problem from turning into a health hazard. In several policy sectors, including the environment, it is difficult to assess the precise impact of public policies.

To avoid the Massachusetts Miracle Fallacy, Putnam suggests focusing on policy outputs rather than on policy outcomes.[36] Borrowed from systems analysis, the concept of policy outputs refers to the products of political institutions – namely, policy inaction or policy actions that generally apply to a sector – whereas policy outcomes refer to the impact of those actions or inactions on the sector.

Following Putnam, I have therefore identified success criteria relevant to policy outputs rather than policy outcomes. In this process, I have assumed that epistemic communities have a central role to play in defining appropriate solutions to policy problems.[37] Taking a broad view of epistemic community, however, I was unable to single out a community of experts capable of providing policy makers with consensual indications as to which agro-environmental policy instruments work and which do not. Therefore, unlike some other studies of policy outputs, including that of Putnam, high-performance environmental policies do not necessarily converge toward a given "one best" policy approach or solution. As explained at length in Chapter 2, epistemic communities in the agro-environmental sector nevertheless agree on three broad principles: (1) the selected instruments should target significant changes in modern farm practices; (2) these changes should concern a wide range of practices; and (3) efforts should be made to spare, to the greatest extent possible, the profitability of agriculture. In light of current knowledge in the agro-environmental sector, I argue that these three principles derive from widely accepted collective values and constitute satisfying external criteria with which to assess environmental policy success in agriculture. In other words, the measure of policy-making performance that I propose in this book rests on the intrusiveness, the comprehensiveness, and the economic sensitivity of agro-environmental policy decisions.

If such policy-making efforts are to be devoted to agricultural pollution, it had better be a serious problem. Information to the contrary would indicate that policy makers are wasting their time and taxpayers' money and that, consequently, they are not deserving of trust. I therefore turn to a discussion of agricultural pollution's seriousness.

How Serious Is Agricultural Pollution in France, the United States, and Canada?

The United States and France are the largest exporters of agricultural commodities in the world, and Canada is not very far behind. To attain such a ranking in terms of agricultural exports, each of the three countries targeted agriculture for protection and state assistance in the postwar period. The idea was that a developed agricultural sector would contribute to broader economic and even strategic policy goals.[38] In turn, state-encouraged increases in production were achieved in the context of a somewhat declining land base in each of the three countries. Intensive agriculture may of course contribute to soil erosion, but urban sprawl is an even more important contributor. In Canada, for example, urban sprawl in the Toronto and Montreal regions has occurred at the cost of some of the most fertile agricultural land in the country. Less land, combined with more production, necessarily engenders higher environmental risks.

In order to remain or to become important exporters of agricultural commodities, farmers in France, the United States, and Canada have had to rely increasingly on inputs such as pesticides and chemical fertilizers for crops, as well as on specialized feed and genetic technologies for livestock. For example, in Canada in 1960, farmers were using an average of 9 kilograms of commercial fertilizer per hectare of arable land; in 1987 the figure was 48 kilograms per hectare. In the US, 39 kilograms of commercial fertilizer per hectare were used by farmers in 1960; in 1987 this figure was 94 kilograms per hectare. In France the situation is even worse: In 1960, 102 kilograms of commercial fertilizer per hectare were used; in 1987 it had reached 299 kilograms per hectare.[39] As a result, yields have substantially increased, especially in France.[40] In 1960 French farmers were harvesting 0.41 metric tonnes of wheat per hectare; in 1996 they harvested 6.5 tonnes per hectare. In the US wheat yields between 1960 and 1996 increased from 0.29 tonnes per hectare to 2.41 tonnes per hectare. The figures for Canada are comparable, with 0.23 tonnes per hectare produced in 1960 and 2.26 tonnes per hectare in 1996.[41] Similarly, the number of days to raise livestock has dropped significantly in all three countries as a result of important increases in farmers' spending on feed and supplements.[42]

Statistics suggest that France has more reasons than the United States and Canada to be concerned about agricultural pollution. This is not entirely surprising given that between the 1970s and 1980s France became the second largest exporter of agricultural commodities, just behind the United States, a country which is many times larger in terms of agricultural land. Furthermore, as the protection provided to French agriculture by the European Common Agricultural Policy is eroded by international trade agreements,[43] competitive pressure appears to be leading to greater farm concentration in the crop and livestock sectors.[44]

To be sure, agriculture was identified as an important source of pollution as early as the 1970s in all three countries. In France, a task force was mandated in the 1970s to study the problem of nitrates in water. This process led to the 1980 Hénin Report, which identifies a number of agricultural practices as contributing to the nitrate problem. In 1972 the United States and Canada began to sign agreements on water pollution in the Great Lakes. In these agreements, agriculture was already associated with water pollution.[45] In 1984 a Canadian Senate report on soil erosion, *Soil at Risk*, made news headlines across the country. Senators complained that if nothing was done about the erosion problem, the agricultural capacity of Canada could be seriously impaired.[46]

Using various techniques, government agencies have more recently quantified the impact of farming on water quality. Recent Environmental Protection Agency (EPA) figures show that agriculture in the United States has contributed to the degradation of 60 percent of the country's rivers and streams that were surveyed.[47] The Institut français de l'environnement showed that 38 percent of the drinking water in France is threatened by agricultural pollution.[48] An Agriculture Canada survey showed that 40 percent of the wells in rural Ontario are polluted at levels above the provincial standards.[49]

Increased agricultural productivity has also caused air pollution problems. Of course, more fossil fuel is needed to run today's modern agricultural machinery. Recent studies even show that gas emissions from livestock and their manure contribute to global warming.[50] In recent years, odours from livestock production have disturbed enough voters to attract the attention of local politicians.[51] In France, the United States, and Canada, rural residents discomforted by strong odours and dust have resorted to public demonstrations to express their concerns about the intensification of agricultural production.[52]

In addition to the general effects of agriculture on the environment, a number of events associated with agricultural pollution have captured media attention in these three countries. Some cases of bacterial contamination of drinking water were particularly serious. The largest accident occurred in Milwaukee, Wisconsin, where about 100 people are believed to have died and 403,000 others were made sick in 1993 after drinking water was contaminated with a parasite called cryptosporidium. The parasite, it has been argued, originated from agricultural runoff.[53] While it was predicted some time ago that cases of illness due to bacteria were likely to appear in Canada's rural areas,[54] a deadly accident finally occurred in Walkerton, Ontario, in the spring of 2000.[55] Bacterial contamination is also a problem in France, but, in addition, cases of drinking water severely polluted with nitrates and agricultural chemicals have been reported.[56] In the spring of 1997, the city of Rennes in France even stopped providing elementary school students with tap water for fear of contamination, notably by atrazine, a pesticide used by corn growers.[57] After reports of problems of nitrate and bacterial contamination of water in France, the OECD issued a "recommendation" for the country regarding agricultural pollution.[58] In Maryland and North Carolina, a microbe called pfiesteria has killed thousands of fish in the past few years and might be associated with human health concerns.[59] There is growing evidence that the outbreak was triggered by livestock manure run-off. One could also mention the broadly publicized lagoon spills in North Carolina in the summer of 1995.[60] According to some accounts, the environmental impact of those spills compares in severity to the Exxon Valdez oil spill, which occurred off the coast of Alaska in 1989.

In short, under the conditions just described, it would be natural to expect policy networks in France, as well as in the United States and Canada, to turn to agricultural pollution and develop policies to address the problem. Failure to do so would constitute proof to the effect that advanced democracies are not properly governed – a situation indicating policy makers might not deserve to be trusted.

A Comparative Research Design

The crisis of confidence in policy makers is widespread within advanced democracies. Putnam, Pharr, and Dalton observe only small differences, distrust appearing only slightly less severe in some Northern European countries.[61] Because the confidence crisis is based on a perception that countries are not properly governed, similarities in distrust levels appear curious since governance structures, or policy networks, markedly differ from country to country. Assuming that governance structures are not all equal, that they encourage differentiated policy-making performances in specific contexts, country variations in confidence levels should be apparent. The failure to observe such variations suggests possible distortions in popular perceptions regarding governance. In any case, this puzzle calls for a comparative research design, with two distinctive features.

First, the selected countries must face similar agro-environmental problems because the nature of problems likely influences policy-making performance. The policy-making perfomance of country A may be superior to that of country B, but little is revealed about the governance capacity of policy networks if policy challenges in country A are much lighter than those in country B. In other words, because the nature of policy problems significantly varies between countries A and B, governance structures do not deserve more trust in the former country than they do in the latter. To highlight the contribution of networks to policy-making performance, it is essential to design a comparative research approach that makes invisible the influence of the nature of problems, and it is for this reason that I have selected countries that face similar agro-environmental challenges.

France, the United States, and Canada have relatively similar agricultural sectors. FAOSTAT, a Food and Agriculture Organization (FAO) database, shows comparable net per capita production index numbers for the year 2001 for the three countries, the United States coming first with 108.4, Canada second with 105.4, and France third with 97.7.[62] Among the top agricultural producers in the world, all three countries have experienced a substantial restructuring of their agriculture since the Second World War, resulting in a significant decrease in the number of farmers and an increase in the size of farms. In the year 2000, as a share of the total population, the agricultural population represented only 3.5 percent in France, 2.6 percent in Canada, and 2.3 percent in the United States.[63] This decline has been accompanied by the intensive use of modern farming practices, requiring significant farm capital endowment. Despite possessing large agricultural economies, France, the United States, and Canada have nevertheless remained agriculturally diversified, combining field crops with the raising of livestock. As discussed above, in each of these countries, farming is sufficiently modern that pollution has become a problem. Although the intensity of intervention using fertilizers and other chemicals is particularly high in France, agricultural pollution also poses a significant challenge in the United States and Canada.

Second, the selected countries should possess different governance structures. If country A and country B face similar policy challenges but are governed by distinctive policy networks, these distinctive networks are likely to enter into an explanation of any differences in policy-making performance. Historical-institutional analysis, in fact, stresses differences in policy-making capacities arising from different institutional contexts.[64] The logic of influence tied to a specific institutional context should shape policy networks in ways that will alleviate or aggravate policy-making performance. When the policy networks of a country are unable to formulate adequate policies, the citizens of this country are justified to complain about governance.

Institutional differences between France, the United States, and Canada are significant enough to expect variations in policy-making performance. France is a unitary country, a form of state whereby subnational governments are subordinated to the central government. In contrast, Canada is a federation in which subfederal governments enjoy wide policy-making autonomy. The United States is also a federation, albeit a more centralized one than Canada. The parliamentary systems of the three countries also significantly vary. France has a semipresidential system of government in which the president exercises tight control over policy making, outside periods of cohabitation. Cohabitation occurs only when the president belongs to a political party different from that of the prime minister, who has to be supported by the Assemblée nationale. In contrast, the American president has little legislative control because Congress operates relatively independently. With a Westminster system of parliament, the Canadian prime minister has more legislative control than the American president. These institutional characteristics of each country are not without consequences for policy networks. The division and the separation of powers that characterize the United States fragment policy networks, some would say, in an exceptional manner. In contrast, the concentration of powers in France, arising from both the form of the state and its semipresidential system, encourages the formation of tight and cohesive networks around those in positions of power. One might however expect these networks to be currently under pressure arising from European integration. Networks in Canada are frequently trapped between the fragmentation logic of federalism and the concentration logic arising from Westminster parliamentarism. Canadian policy networks are likely to vary extensively depending on whether they operate in federal, provincial, or shared sectors of jurisdiction.

Policy studies on agriculture have documented the differences between the three countries at the level of policy networks. In France, agricultural policy making has been handled through a corporatist policy network,[65] whereas in the United States, Congress has often played a determinant role in a network otherwise increasingly fragmented.[66] Unsurprisingly, agricultural policy networks in Canada are relatively segregated between the provincial and federal levels of government, with stronger networks located in the provinces.[67] Because such differences exist, one might expect policy-making performance in the agro-environmental sector to vary from one country to the next. In turn, such variations should raise questions as to whether the crisis of confidence deserves to be as widespread as it is.

Lastly, it should be noted that the research efforts leading to this book concentrated on the agenda-setting and policy-formulation stages of the policy-making cycle. Questions related to the policy-implementation and policy-evaluation stages were largely left out. The information on agenda setting and policy formulation in France, the United States, and Canada was primarily obtained through nearly one hundred confidential interviews conducted between 1997 and 1998 with officials representing all the major organizations involved in agro-environmental policy. Several of these interviews were recorded and fully transcribed. Quotes that did not risk breaching the confidentiality of the interviewees are reproduced in Chapters 4, 5, and 6. The information obtained from these interviews was verified, to the greatest extent possible, using official documents and secondary sources. Official documents were also utilized to update the analysis to the spring of 2002.

The Book's Outline

Overall, I argue in this book that distrust in governance arrangements is misplaced on two counts. First, the empirical evidence suggests that, more often than not, policy networks do in fact adequately address serious problems. While performance levels appear to be higher in France and in the United States, no network associated with agriculture and the environment in any of the three countries regarded agricultural pollution as a problem undeserving of its attention. Second, the factors usually suspected for policy-making failures were not found to have such effects in France, the United States, and Canada. In fact, there is no reason to suspect that bureaucracy, interest groups, or internationalization necessarily engender poor policy-making performance. Rather, it was found that under specific conditions, each of these factors can have positive effects on performance levels. In other words, beliefs to the effect that policy networks generally fail to address problems adequately are unjustified. The next six chapters will attempt to make this argument a convincing one.

In Chapter 2, I present a method for assessing policy-making performance in the agro-environmental sector. After discussing the objective- and solution-oriented methods, I propose an assessment based on points of convergence between otherwise competing epistemic communities. Three competing agro-environmental epistemic communities are identified. I show that despite competition over the specific manner of approaching agricultural pollution, these communities are in relative agreement over three policy beliefs: (1) environmental policy must bring major changes in farming practices; (2) it must cover a comprehensive range of farming practices; and (3) it should be economically sensitive. I thus conclude that highly proficient agro-environmental policy-making performance occurs when it allows the development of policies consistent with these beliefs.

In Chapter 3, after a brief discussion of agenda setting, I present the network approach, which informs the central argument of the book. I argue that problem solving is most likely to occur when actor constellations properly balance cohesion and diversity. In addition, network structures that attribute a central role to civil society actors, but that distribute power evenly between them and state actors, are those likely to yield the highest performance because, following Weiss, they enable "governed interdependence."[68] I then move on to discuss a number of policy-making theories suggesting that real-life networks are unlikely to be of this nature. In fact, theories on the new politics of the welfare state, on regionalization, and on internationalization appear to constitute solid foundations for the crisis of confidence.

However, these latter theoretical proposals are quickly dismissed in Chapter 4, the first of three empirical chapters. Chapter 4 shows that France, between the 1980s and 1990s, moved from poor to high policy-making performance. It is shown that multilevel governance, which strengthened environmental bureaucracy without threatening interest groups' role within a corporatist network, did not prevent this change in performance level but rather enabled it to occur.

Chapter 5 shows that although American policy makers did not attain a performance level as satisfying as that found in France, they still obtained good results. In fact, the central problem is one of coordination between federal and state agro-environmental policy. This lack of coordination is attributable to the presence of two complementary but largely autonomous actor networks rather than, as theoretically suspected in Chapter 3, to bureaucracy, interest group politics, or internationalization.

Chapter 6 serves to confirm that these factors are not major causes of bad policy-making performance. Of the three countries, Canada displays the least satisfying performance level, thereby justifying citizens' fears that environmental policy makers fail to live up to their promises. However, citizens would be targeting the wrong causes if they were to blame bureaucrats and self-interested organized groups. In fact, it seems that only better group organization and stronger bureaucracies within environmental policy networks would lead to an improvement in Canada's environmental policy-making performance.

After a systematic comparison of these three countries, Chapter 7 revisits the distrust-inspiring theories presented in Chapter 3. Central to this chapter is the demonstration that these theories not only unjustifiably nourish the crisis of confidence, but also misplace the responsibility for policy-making failures. While the theories are not rejected altogether, adjustments are proposed to give them a trust-inspiring twist.

2
Assessing Policy-Making Performance

Policy makers are commonly accused of being incapable of dealing with today's problems. The public tends to be rather severe, often failing to consider the difficulty of the task of resolving most collective problems and preferring to blame politics or governance arrangements. Some critics allege that global politics has reduced the number of alternatives available to tackle problems, whereas others blame state building, arguing that it has encouraged the proliferation of interest groups that divert the attention of policy makers toward distributive issues and away from real problem solving. Still others explain the incapacity of policy makers by suggesting that competent people will choose the dynamism of private-sector employment over the rigidity of public-sector work. Such opinions are necessarily based on some sort of assessment of governance, but these assessments are often questionable.

Indeed, popular assessments of policy networks' capacity to solve collective problems are often intuitive or based on anecdotal evidence. Journalists, for instance, often choose a policy and present it in such a way as to illustrate the presumed incompetence of public office holders. Naturally, all policy evaluations are not conducted with such total absence of rigour; it is now increasingly common for the public to be exposed to professional policy evaluations. However, even these policy evaluations must be considered with caution, although their alleged scientific neutrality may inspire confidence. Policy evaluations, even when conducted by professionals, cannot escape politics.[1] They have notably been used by neoliberal think-tanks to service the political objectives of state retrenchment. The technique is easy: Success is constructed so narrowly that high performance can be attained only under exceptional circumstances. Hidden behind a complex jargon, these policy evaluations rarely allow the lay public to question the normative choices that lie behind them.[2]

This being said, I cannot claim to propose a method allowing for accurate and neutral assessments of policy-making performance. Rather, the objective of this chapter is to present as clearly as possible one cohesive strategy, anchored in collective values, for assessing policy decisions in a particular sector. Assuming that there is no universal and optimal policy evaluation method, I stress the importance of designing a strategy capable of capturing the specificity of the policy problem examined: namely, agricultural pollution. But I begin with a review of existing methods.

Approaches to Policy Evaluation

As Putnam points out: "Appraisals of institutional performance are rare."[3] Few political scientists have ventured to explore the normative question of how well governments generally perform their job. This is symptomatic of the difficulty of producing such assessments, and it unfortunately leaves citizens with little more than anecdotes and partial evaluations on which to base their opinions of policy makers. Sector evaluations nevertheless present a certain potential since their aggregation could likely produce a general sense of government performance. However, carrying out such an aggregation in a rigorous manner is fraught with serious difficulties. Few sector policy evaluations are comparable since authors rely on different concepts, measures, and methodologies to produce them. Aggregation possibilities are consequently reduced to such an extent that any general assessments of state performance produced in this manner would be unconvincing at best.

As explained in the introduction, a problem-based approach is promising when performance faces political obstacles significant enough that success can only inspire a general sense of trust in policy makers. I should stress that the objective of the evaluation method presented in this chapter is not to generalize from a single case but rather to allow for the formulation of an inspiring, rigorously constructed story. Before turning to this method, I will present two alternatives for problem-based evaluations.

The Objective-Oriented Approach

The evaluation of policies targeted at clearly defined problems has the advantage of being approachable in terms of objectives. The objective-oriented approach to policy evaluation uses objectives as benchmarks. The performance of policy makers is satisfactory to the extent that they meet the objectives that were set. Whereas the general purpose of government is often the subject of lively debates, sectorial objectives usually go unchallenged. For example, no one would question that the objective of economic policy is to stimulate growth and employment, and that the production of economic policy evaluations is facilitated by the presence of these measurable objectives. Government performance is thus considered satisfactory to the extent that policy makers are able to stimulate growth and lower unemployment.[4]

Several problems besides economics have been the object of evaluations using the objective-oriented approach. Steinmo, for example, uses this approach for taxation, where the objective – that is, revenue generation – is rather straightforward and uncontested. In addition, revenue generation, just like economic growth and unemployment, is easy to measure and compare. Using data on the revenues generated by tax systems, Steinmo concludes that Swedish performance is higher than British performance, which is in turn higher than that of the Americans.[5]

However, government objectives are not always this easy to identify and may vary frequently over time, making comparisons with results a hazardous task. For several problems, and notably environmental problems, a certain amount of time is necessary between the setting of policy objectives and the appearance of tangible results. Poor environmental results may thus be less the consequence of policy failure than the consequence of an evaluation conducted too early. In addition, Pal stresses that the general and uncontroversial goals associated with most policy problems are often located at a level of abstraction too high for policy making. The objects of most policies are specific – that is, they generate debates and reveal severe disagreements.[6] For example, the general objective of taxation might be to raise revenues, but the objective of a tax on gas may be to encourage alternative technologies and consumption of alternative sources of energy. Debates and disagreements on the appropriateness of stimulating the emergence of innovative technologies at the expense of the oil industry would encourage governments to remain vague about the precise objective of such a tax. Hence an evaluation of a tax on gas based on the objective-oriented approach may prove to be problematic.

Even where objectives are clearly stated and early results can be expected, deciding whether results meet objectives depends on interpretation: A partially attained objective can always be viewed either as a glass half full or as a glass half empty. While interpretation cannot be entirely eliminated from policy evaluation, comparative research designs allow for the presentation of assessments in relative rather than absolute terms: The glass half full or half empty becomes country Y doing better than country X.

However, environmental data are difficult to compare from one country to the next. The Organization for Economic Cooperation and Development (OECD) encourages governments to harmonize environmental data collection, but data collection remains the responsibility of individual countries, which must balance the demands for harmonization against their own concerns.[7] Collecting data on environmental quality is a process contingent on geography and on the nature of the polluting activities. Assessing water quality, for example, involves decisions on which rivers to sample, the sampling sites for each river, and the frequency of sampling –

decisions highly contingent both on climate and on the nature of the rivers and the land through which they flow. Similarly, decisions concerning substances to be tested are based on available knowledge about polluting activities. The wide range of possible contaminants and the costs of water analyses require that tests be designed to focus on those toxic substances most likely to be present in large quantity.[8] In addition, these difficulties in harmonizing environmental data collection underscore the political conflicts that may arise in countries where upgrading is necessary since not all social groups prefer thorough environmental data. In any case, studies comparing environmental quality have generally been accompanied by serious warnings about their validity.[9]

Even more problematic in this sector is the diversity of influences on environmental quality. Vogel argues that it is nearly impossible to distinguish from other factors the impact of public policy on environmental improvement or degradation[10] despite the availability of time-series data sets in many countries and sophisticated statistical techniques. Economic development, both qualitatively and quantitatively, certainly has a far greater influence on environmental quality than does public policy.

The Solution-Oriented Approach

Studies of performance in the environmental sector have thus focused on the policy process and asked whether it yields plausible solutions to problems. Here, countries display a high problem-solving capacity when they converge toward policy instrument types believed to be optimal. At least seven instrument types can be identified in the environmental sector. These are listed in Table 2.1. As an illustration of the functioning of the solution-oriented approach, American economists claim that command-and-control regulations in the environmental sector have proven to be inefficient and need to be replaced by economic instruments such as tradable permits or deposit-refund systems.[11] Consequently, they present the failure of the American government, or of any government for that matter, to move decisively in this direction as an indication of the deficiency of policy making.[12]

However, the question arises as to how one comes to decide that an environmental policy instrument type will provide a plausible solution to a problem. It is clear that even though the economic approach to environmental problems has gained in popularity, there is no consensus around this approach. Moreover, if policy evaluations view success as the adoption of a given policy option, we ought to be absolutely certain that the given option is the right one. Perhaps the closest we can come to such certainty is when an expert consensus emerges over a given issue.

In the environmental sector, the literature on epistemic communities

Table 2.1

Environmental policy instrument types: Definitions, characteristics, and degree of epistemic community approval

Instrument types	Definitions	Epistemic community approval		
		Ecological scientists	Agriculture system analysts	Alternative agriculture
Moral suasion/education	Education to sensitize farmers about the environmental impacts of their practices	×	√	–
Economic incentive/subsidy	Provision of economic incentives to encourage the adoption of given environmental practices	o	√	–
Command-and-control regulations	Adoption of coercive rules to prevent reliance on polluting farm practices	√	×	–
Cross-compliance	Transformation of compliance with environmental standards into a condition of eligibility for regular farm programs	√	×	×
Endogenous	Promotion of the environment as an asset that can contribute to improving the economic situation of farmers	o	o	√
Reform	Seeks a redefinition of farming, away from the objective of always producing more	o	×	√
Self-governance/self-regulation	Leaves to polluters the entire responsibility to decide which practices are acceptable and which are not	×	o	o

√ = strong approval, – = some approval, × = strong disapproval, o = no opinion draws attention to the formation of expert consensus. Epistemic communities are networks of individuals whose common values engage them in the generation of specialized knowledge through a shared scientific paradigm.[13] While heated debates take place within any epistemic community, the presence of a common paradigm equips them to construct consensual solutions for the problems they are concerned about. However, this literature, grounded in the history of science, displays no blind faith in the capacity of these communities to produce correct advice.[14] As Majone points out, the policy process is more about crafting arguments and rationales than about finding fail-safe solutions.[15] Because they produce arguments and rationales, epistemic communities, it is argued, provide welcome policy-making guidance in complex sectors where policy makers face uncertainty in the designing of policy.[16] Their unity, their roots in rigorous academic disciplines, and their detachment from economic gains all serve to inspire trust.

It is important at this stage to insist on the distinction between epistemic communities and policy networks. Haas aptly insists that members of epistemic communities exert influence on policy most effectively when they can participate directly in policy networks,[17] but the functions of both entities are distinctive. Policy networks make public policy; "epistemic communities are channels through which new ideas circulate."[18] Epistemic communities' essential role is the production and diffusion of policy ideas, while policy networks' essential role is choosing among these ideas. Some members of epistemic communities can decide to integrate with policy networks to circulate their ideas more efficiently, but several of them will continue operating from outside.

In any case, in his study of Mediterranean pollution control between 1972 and the early 1980s, Haas found "a publicly recognized group with an unchallenged claim to understanding the technical nature of the ... substantive issue-area ... able to interpret for traditional decision makers facts or events in new ways."[19] The group stressed the need for greater cooperation between the Mediterranean governments to devise "rational forms of economic planning to internalize environmental considerations into virtually all forms of policymaking."[20] The weight given to the ideas associated with this community enabled the integration of members into different government functions, thereby guaranteeing a high level of compliance and success in cleaning up the Mediterranean.

Similarly, Rabe and Zimmerman see the compliance of governments with the ideas of an environmental epistemic community as a sign of success. One North American epistemic community claims that the Great Lakes basin is particularly vulnerable to cross-media pollution – that is, to air and soil pollution transformed over time into water pollution. The integration of members of this community into state and provincial networks in the Great Lakes area has enabled the devising of integrated regulatory approaches that minimize the transport of pollution through air and soil from one jurisdiction to another.[21]

While such an evaluation of government policy success appears unproblematic, the conditions met in the studies both of Haas and of Rabe and Zimmerman are not frequently encountered. Besides having in common the view of success as compliance with the ideas of an epistemic community, the studies share a focus on a limited geographic area. Obviously, geographic proximity can contribute in several ways to the creation of unchallenged epistemic communities. Proximity enables the frequent meetings required for engaging in dialogues likely to encourage convergence around a scientific paradigm and, eventually, around common policy positions. Likewise, proximity contributes to focusing dialogues around specific problems, such as pollution in the Mediterranean (Haas) or in the Great Lakes (Rabe and Zimmerman). Such a focus reduces the likelihood of disjointed communication and fosters the mobilization of epistemic community members. However, for a general assessment of government performance, it may be desirable to encompass larger geographic areas. But, at the level of a polity, the presence of unchallenged epistemic communities is compromised.

As suggested earlier, no epistemic community has imposed itself in the debate on the type of policy instruments governments should generally favour to protect the environment. The idea that economic instruments are superior to command-and-control regulatory instruments may have gained popularity,[22] but it certainly does not go unchallenged. As Majone argues: "Cautious conclusion seems all the more appropriate in view of a recent survey indicating that even among economists the proportion of those who are convinced of the superiority of charges [a specific economic instrument] is much smaller than one would assume from the near unanimous endorsement by textbooks on environmental economics."[23] Concurrent ideas concerning ways of tackling agricultural pollution in fact appear to arise from three epistemic communities.

The first is a community of scientists who, as discussed by Haas, believe that economic growth or other policy objectives should be subordinated to protection of the environment. It has been documented that the agricultural sector, since the end of the Second World War, has enjoyed exceptional treatment.[24] Briefly stated, policy makers were inclined to subordinate most policy objectives, including protection of the environment, to the development of the agricultural sector.[25] With mounting evidence that the agricultural practices associated with the first green revolution had a significant environmental impact, a community of ecological scientists, biologists, chemists, health scientists, and others began requesting the end of agricultural exceptionalism with respect to the environment.

The essence of their demands was that agriculture be subject to regulations as strict as those enforced in the point source pollution sectors. The influence of the members of this epistemic community is significant, as several of them work in environmental ministries across OECD countries.

More recently, an epistemic community of agricultural system analysts has emerged to challenge the community of ecological scientists. This community of agricultural system analysts is also an interdisciplinary community drawing on a number of agricultural sciences, including agricultural economics. It is held together by a scientific paradigm that views agriculture as a *system* whereby nature is domesticated for the purpose of producing food.[26] Agricultural system analysts are further unified in the belief that the primary function of these systems – that is, food production – constitutes a vital function.[27] As a result, the members of this epistemic community subordinate all objectives, including environmental protection, to the efficacy of agricultural systems. Unlike ecological scientists, agricultural system analysts tend to shy away from producing agro-environmental policy advice that significantly constrains agricultural production (see Table 2.1).

This is not to say that environmental protection does not constitute an important concern for agricultural system analysts. In fact, they have increasingly given attention to the problems of agricultural pollution as they have grown concerned about the impact of resource depletion on food production.[28] As one analyst of agricultural systems put it: "The technology selected for agricultural production, characterized by levels of inputs and cultivation practices, not only determines agricultural output, but also affects the quality of the soil and water. For example, some soil erosion and changes in soil chemistry are usually associated with the production of a crop. The future productivity of soil is thus affected."[29]

Here, it may be important to distinguish between the policy advice emerging out of agricultural sciences and the advice formulated by agricultural economists. Agricultural scientists tend to stress the importance of generating knowledge on sustainable farm practices and diffusing the information to farmers. Informed farmers should naturally favour practices that prevent the depletion of the resources upon which they depend.[30] In contrast, agricultural economists argue that under certain conditions informed farmers should not adopt sustainable practices. This would notably be the case under conditions of imperfect competition – that is, when prices do not reflect the value of adopting new technologies and practices.[31] They also state that farmers should limit the adoption of new practices to areas where they have to individually bear the full cost of pollution or where market failures exist.[32] These conditions may warrant stronger government intervention than the agricultural scientists are suggesting. In formulating advice for these interventions, agricultural economists have stressed the importance of the diffused or nonpoint source nature of agricultural pollution.[33] Given the number and geographic dispersal of farms, it is very difficult to trace the exact origin of agricultural pollution – a situation that renders the use of traditional command-and-control environmental regulations problematic. The monitoring and enforcement costs in the agricultural sector are presumed to be too high to justify this type of policy instrument. Agricultural economists have thus prescribed the use of economic instruments, including subsidies, charges, taxes on pollution, and markets in pollution rights.[34]

However, the political consequences of any differences in policy advice between agricultural scientists and agricultural economists should not be exaggerated. As Haas argues, epistemic communities are conducive to developing policies that are satisfying to all members without direct confrontation.[35] Advice from the agricultural system community tends to be presented to policy makers as a unified set of proposals that challenge those of the ecological scientists. Moreover, the policy ideas of agricultural system analysts have attracted serious attention in recent years, being assimilated by powerful ministries of agriculture.

Adhering to a narrow definition of expert knowledge, some observers might hesitate to associate alternative agriculture with an epistemic community, even if critical theory invites analysts to broaden their perspective on knowledge.[36] I argue that much of the knowledge on alternative agriculture emanates from an epistemic community because experts working from this perspective invoke science and adhere to causal beliefs. Although more marginal, this epistemic community poses a serious challenge to both the ecological scientists and agricultural systems communities (see Table 2.1).

Proponents of alternative agriculture view farming as a *way of life* in harmony with nature and rural communities. In contrast to agricultural system analysts, they believe that agriculture fulfils ecological and social functions in addition to ensuring food production. They criticize the modernization of the sector for having created an imbalance between these three functions in favour of the latter.[37] The alternative agriculture community proposes nothing less than a reform of modern agricultural practices. Members of this community, largely concentrated in Europe, often practice agriculture but also include social scientists interested in the impact of farming practices on the environment and society. Unlike the two previous communities, which are structured around typical academic forums, the alternative agriculture community is structured around groups such as France's Centre d'étude pour un développement agricole plus autonome (CEDAPA), which does not wholeheartedly embrace positivist methods. Perhaps it is for this reason that the alternative agriculture community, more popular in Europe than in North America, remains a marginal one.[38]

Nevertheless, the coexistence of three epistemic communities pertaining to worldwide agricultural pollution is problematic for the utilization of the solution-oriented approach to policy evaluation. Science-guided policy making is contestable on many grounds,[39] but it at least has the merit of suggesting a basis for action where uncertainty and inaction otherwise prevail. However, the uncertainty resumes when epistemic communities become numerous and start to challenge each other, as is the case in the agro-environmental sector. Under these circumstances, policy evaluations cannot convincingly construct success as convergence toward the preferred policy instrument type of a single epistemic community. Any competition among epistemic communities indicates an absence of consensus among experts. Agricultural system analysts may have gained popularity in recent years, but the continuing relevance of ecological scientists, and the increasing relevance of scientists advocating alternative agriculture, prevents use of the instruments proposed by the first community as indicators of success. A policy evaluator who decides, for methodological reasons, that success is attained when convergence toward the policy instruments proposed by agricultural system analysts occurs is in fact deciding that the environment should be subordinate to agriculture. It is my view that this is not a decision policy evaluators should be making.

Constructing an Evaluation Method for Agro-Environmental Policy Making

It is also my view that the solution-oriented method should not be rejected altogether. The evaluation proposed in this book simply takes one step backward from this method. Again, epistemic communities are considered useful when they convey a high level of certainty that a given package of policy instruments constitutes a plausible solution to a problem. Naturally, the presence of three competing epistemic communities in the agro-environmental sector prevents the formation of an expert consensus on an adequate response to agricultural pollution. But this focus on the end result of consensus formation overshadows ideas that needed to be addressed at earlier stages in the process of epistemic community formation, ideas located at a higher level of abstraction. It is interesting to note that the level of agreement among the three epistemic communities on these ideas – and even among a vast lay public – can be quite high. Before proposing direct regulations, economic incentives, or an alternative agriculture, experts had to agree that farming could cause pollution and that those practices potentially damaging to the environment needed to be changed. Thus epistemic communities might not offer certainty as to which agro-environmental policy instrument type should be endorsed, but they leave little doubt that agricultural pollution is serious enough to justify efforts to protect the environment of rural areas. The literature often highlights the fact that public policy studies should pay as much attention to what governments fail to do as to what they actually do.[40] And, indeed, given the evidence produced by the epistemic communities, a failure on the part of governments to adopt any instrument of one type or another appears significant. In the face of such a failure, governments can hardly be said to contribute to problem solving. Epistemic communities can therefore still be useful in constructing a method for assessing policy-making performance.

Policy evaluation needs to be detached from the rationales generated by the epistemic communities in producing specific advice on policy instruments and needs to draw on more encompassing rationales that may shape policy decisions without dictating convergence on any instrument type. These types of rationales often obtain support across epistemic communities and beyond, thereby contributing to the reduction of uncertainty. Moreover, stepping away from the specific policy advice of environmental specialists leaves ample space for discussion of scientifically irresolvable goal and belief conflicts[41] among interested citizens and eventually for the adaptation of policy alternatives to local circumstances. We saw that all three epistemic communities, to stress the necessity of adopting agro-environmental policy initiatives, use evidence of agricultural pollution in the countryside as a rationale. While reducing uncertainty with regard to the necessity to act on agricultural pollution, this rationale is not authoritative enough to discourage dialogues about whether environmental protection is inconsistent with the profitability of agriculture, about which practice pollutes most, or about which package of instruments should be adopted. I have identified three additional rationales that, in a similar fashion, cut across agro-environmental epistemic communities and present a potential for reducing uncertainty without discouraging the participation of nonexperts in agro-environmental decision making.

The first of these rationales builds on the idea that changes in agricultural practices are not equally significant. For example, experts would agree that planting more trees around farm buildings embellishes the landscape but is not as significant an environmental practice as adequately managing farm animal wastes. Good waste management, unlike farm embellishment, is not a precise practice. Furthermore, waste management is more likely to have a real impact on the resources (air, soil, and water) threatened by agricultural pollution. However, there are disagreements as to what constitutes appropriate waste management practices. Agricultural system analysts, for example, would prefer to target management practices that not only protect the environment, but also protect soil productivity. But the suggestion that waste management practices are more significant than farm embellishments does not enter into this debate. It is self-evident that the adoption of any new waste management practice is a more demanding behavioural change than simply planting new trees. In fact, the central element of this rationale is the degree of change required in farm practices – what I call "intrusiveness." All three epistemic communities agree that environmental protection of rural areas requires the use of policy instruments intrusive enough to engender significant changes in the behaviour of farmers.

Suggesting that a policy must be intrusive in order to produce significant results is different from advocating the adoption of a specific type of policy instrument. For example, both command-and-control regulations and subsidies can be intrusive. Subsidies are intrusive to the extent that they are targeted at significant environmental practices and are generous enough to incite their adoption. Likewise, regulations are intrusive when they target particularly damaging practices and severely sanction violations. However, by nature, moral suasion and education are almost always nonintrusive. While an aggressive publicity campaign can be viewed as intrusive, education will always remain less intrusive than command-and-control regulations because of the voluntary character of the requested behavioural change.

The second rationale emphasizes the importance of taking the problem of cross-media pollution into account.[42] Cross-media pollution is indeed a rather acute problem in agriculture. Obviously, agricultural use takes up a significant proportion of land, which, if polluted, becomes an efficient vehicle for ground and surface water pollution. At the same time, certain manure treatment technologies that prevent soil and water pollution may release significant amounts of atmospheric pollutants. Ecological scientists and members of the alternative agriculture epistemic community adhere rather tightly to this rationale, which demands comprehensive policies. A comprehensive package of policy instruments, notwithstanding their particular type, covers a wide range of practices, thereby preventing the pollution of air, soil, and water rather than the pollution of only one of these media.

Agricultural system analysts are not quite as committed to comprehensiveness but do not altogether disagree. As discussed above, agricultural scientists tend to see pollution as a problem to the extent that it threatens agricultural productivity, and water and soil pollution certainly poses such a threat. Crop yields and livestock health both depend on good water and soil quality. On the other hand, the impact of air quality on agricultural production is not quite so direct. Several agricultural economists nevertheless stress that "the producer should use the environment as a production factor up to the limit at which his marginal production cost equals the cost to society of the marginal unit of pollution."[43] Presumably, then, the cost to society of environmental utilization by farmers includes air pollution. In short, there is a difference in commitment to comprehensiveness between agricultural system analysts, on the one side, and ecological scientists and proponents of alternative agriculture, on the other, but this gap can be bridged.

The final rationale, which also suffers from a similar minor disagreement between epistemic communities, pertains to the economic impact of changing agricultural practices to protect the environment. Briefly stated, any policy is better when it achieves its objectives without harming the economy.[44] The discourse on sustainable development is rather optimistic about the possibilities of meeting environmental objectives under conditions of economic growth. Without sharing this optimism, the third rationale stresses the need to develop environmental policies that do not disregard the viability of economic activities, especially agriculture. The idea here is not to perpetuate agricultural exceptionalism[45] but to promote policies that seek to attain an equilibrium between environmental protection and agriculture's economic viability.

From the point of view of proponents of alternative agriculture, this rationale is unproblematic. Their discourse begins with the observation that current policies are in a situation of disequilibrium favouring the economic axis. They claim that a better balance between the economic, environmental, and social functions of agriculture needs to be attained.[46] Most agricultural system analysts would share this view that policies are too often geared toward production without proper attention to resource depletion. While unwilling to go as far as the proponents of alternative agriculture on the environmental function of agriculture and the ways public policy should account for it, they demand that an equilibrium be established between agricultural development and environmental protection.

The minor disagreement over this rationale originates within the ranks of the community of ecological scientists. Several of these scientists claim that the attainment of an equilibrium between environmental objectives and economic growth is possible only under exceptional circumstances. They often tend to understand environmental protection and economic growth in mutually exclusive terms.[47] Given the values to which they adhere, they claim the environment should prevail over the economy.

This disagreement is minor in that the objections of ecological scientists are general and tend to disregard the unique characteristics of agriculture. As the proponents of alternative agriculture point out, farming, unlike other economic activities, fulfils some environmental functions that ecological scientists might not wish to see abandoned. Farmers have traditionally fulfilled the function of steward of the rural landscape. Consequently, the choice is not so much between the environment and the economy as between certain forms of environmental protection and others that incidentally depend on the viability of agriculture. In addition, agriculture is a vital activity to humans. This is particularly true of the three countries analyzed in this book. France, the United States, and Canada make a net contribution to food provision outside of their borders. Under these circumstances, it is reasonable to expect French, American, and Canadian policy makers to attempt to achieve equilibrium between agriculture's economic viability and environmental protection, even if they cannot actually achieve it.

It would be useful at this point to underline that threats to the economic viability of agriculture may on occasion be less a matter of aggressive environmental policy making than of bad policy instrument coordination. As stressed in the literature, the problem of environmental policy coordination may be particularly acute in federal systems.[48] As illustrated in Chapter 5, American policy makers are concerned with the economic viability of agriculture and thus have developed economically sensitive packages of policy instruments. However, the absence of coordination between state command-and-control regulations and federal financial aids certainly threatens a number of farmers. The assessment of the economic sensitivity of an instrument package thus presents significant challenges, not least of which is the requirement to account for the coordination of instruments' distributing costs and benefits. I return to this issue in Chapter 5.

Table 2.2 provides an overview of the level of agreement across the three epistemic communities on the characteristics of agro-environmental policy and their underlying rationales. As presented in the previous section, the low level of agreement among the three epistemic communities on instrument types creates some uncertainty as to whether policy makers should opt for command-and-control regulations, economic incentives, moral suasion, or other instruments. However, epistemic communities adhere sufficiently to rationales more-or-less removed from any specific advice on instrument type to convey relative certainty on a number of characteristics of agro-environmental policy. First, policy makers should be in no doubt that agricultural pollution constitutes a problem needing their attention. Likewise, policy makers should design policy instruments intrusive enough to induce significant changes in farming practices. These instruments, a third rationale suggests, should be packaged to cover a range of agricultural practices comprehensive enough to avoid cross-media pollution. And, lastly, policy makers should have little doubt that the economic viability of agriculture needs to be considered during this packaging process.

These characteristics of agro-environmental policy, which are the object of a consensus, form the basis of the evaluation method employed in this book. Government performance depends not on convergence toward one type of instrument or another but on the extent to which the packages of instruments display the various characteristics that are the object of

Table 2.2

Level of agreement across three epistemic communities on the characteristics of agro-environmental policy

Epistemic community	Necessity of action	Intrusiveness	Comprehensiveness	Economic viability	Instrument type
Ecological scientists	√	√	√	–	×
Agricultural system analysts	√	√	–	√	×
Alternative agriculture	√	√	√	√	×

√ = agreement, – = slight agreement, × = strong disagreement consensus among the three epistemic communities. It is interesting to note that this departure from a method tending to view convergence toward one type of instrument or another is not entirely inconsistent with the environmental policy literature. Lotspeich argues that "the relevant choice is not between the two [environmental] approaches, but rather one of the correct mix of market and CAC [command-and-control] instruments."[49] Writing more specifically about agro-environmental policy, Weersink and his colleagues note that "the optimal strategy for any given situation will likely involve a mix of instruments. Economic instruments could be used in conjunction with the two other major environmental policy choices, moral suasion and direct regulation."[50] Under these circumstances, the disagreement between epistemic communities over instrument type should not be too alarming to policy makers, as they would be well advised not to rely on a single one.

Conclusion

This chapter provides the necessary conceptual material to conduct assessments of policy-making performance in the agro-environmental sector. The method, it should be emphasized, is not universal, but provides tools for evaluating a sector that possesses a significant number of unique characteristics. The contribution of this chapter to the subdiscipline of policy evaluation emphasizes the nature of the problem at which policy solutions are aimed. Given the specificity of policy problems, not to mention the diversity of collective values, ready-made methods, it is argued, cannot be universally applied.

Nevertheless, in the agro-environmental sector, I have shown that one can find inspiration in the solution-oriented method, even though the absence of an expert consensus requires some creativity. I have suggested that taking a step backward from specific policy advice in processes of expert consensus building is a promising avenue. In fact, I have shown that three epistemic communities share more-or-less similar ideas in early stages of their formation processes. They agree that agricultural pollution needs to be the object of policies; they agree that these policies have to be intrusive – that is, require major changes in agricultural practices; they also agree that these policies should be comprehensive – that is, cover a wide range of practices; and lastly they agree that the economic viability of agriculture should be considered in the process of designing policy. Relative certainty about the appropriateness of these policy characteristics thus emerges out of the epistemic communities, and consequently I feel safe in using them as indications of good performance. These policy characteristics constitute the safest obtainable indications that policy decisions in the agro-environmental sector have real problem-solving potential. One should therefore normally expect policy decisions that possess these characteristics to contribute to the output-oriented legitimacy of modern states.

What kind of policy networks would plausibly possess the capacity to adopt environmental policies endowed with these characteristics? Should we expect real-life policy networks to be fraught with so many deficiencies as to have significant difficulties producing such agro-environmental policies? I turn to these questions in the next chapter.

3
Networks and Performance

What type of governance structure would, in theory, allow the designing of intrusive, comprehensive, and yet economically viable environmental policies for the agricultural sector? Drawing on the policy network and agenda-setting literature, I attempt to imagine in this chapter the ideal governance structure – that is, the structure most likely to achieve the expected results in the agro-environmental sector. I also show how efforts to comprehend real-life governance unfortunately leave little hope that policy networks will even approximate this ideal structure.

Popular assessments of policy-making performance are influenced by theoretical constructions, several of which claim that governance arrangements suffer from serious defects. Sophisticated theories present current political conditions in such a manner as to inspire distrust in the capacity of policy networks to design adequate public policies. The state has grown considerably in size since the Second World War, and consequently public bureaucracies have become large organizations that present sizable challenges to managers. These bureaucracies are often pictured as being engaged in a competition resulting in duplications and accumulations of responsibilities that create diversions from their original mandates. Responsibility increases, in turn, can translate into new channels of access to the state for resource-hungry interest groups not content with traditional democratic processes. Globalization adds to these difficulties, as policy makers must now contend with transnational interests. In addition to being accountable to domestic constituencies, policy makers must now respond to international demands.

In this chapter, I specifically examine agenda setting, theories relating to the new politics of the welfare state, and theories stressing the impacts from internationalization and regional integration to show that the confidence crisis finds a significant source of encouragement in policy-making theories. In fact, these representations of the hindrances to performance-generating network conditions inspire nothing but suspicion toward governance.

Agenda-Setting Theories and Policy-Making Performance

Among the few certainties emerging out of epistemic communities is the necessity for policy makers to act on agricultural pollution. Scientists and experts might not agree on an appropriate environmental policy for agriculture, but they agree that the problem of agricultural pollution should not be left unaddressed. Of course, agricultural pollution has to first get onto the political agendas of governments for any action to be taken.

Agenda setting is defined as the process whereby collective problems also become understood as political problems – that is, problems of concern to governments. The most straightforward theory of agenda setting asserts that problems become political when they become really serious. Some authors, for example, have argued that economic development in OECD countries during the twentieth century gave rise to sufficiently serious social problems that the welfare state had to emerge.[1] Likewise, the development of the agricultural sector may be seen as having given rise to environmental problems so serious in nature that policy makers could no longer simply ignore them. Given the serious character of agro-environmental problems that have been brought to light by experts from various backgrounds, agricultural pollution should end up more-or-less automatically on the political agendas of countries where modern agriculture is an important economic activity. This manner of understanding agenda setting is certainly reassuring from the point of view of policy-making performance.

However, competing theories of agenda setting lead to different expectations. Kingdon elaborates the most influential of these theories. Drawing on the garbage can model of decision making, he argues that in any political system, a number of policy entrepreneurs work toward getting their ideas onto the political agenda.[2] They do so by taking advantage of the coupling of policy ideas with emerging problems when political circumstances are favourable. This coupling, however, is far from being a straightforward linear process. First, policy ideas are shaped through combination and recombination into what Kingdon calls a policy primeval soup – a process that bears some resemblance to certain biological processes.[3] Second, a political stream, independent of the stream of policy ideas, comprises factors such as the election of a new government, interest group lobbying, or, more simply, the perception held by politicians of the changing national mood.[4] Third, a problem stream functions as a channel for the problems of society.[5] Policy entrepreneurs are presented with windows of opportunity only when these three independent streams come together.

The capacity of policy entrepreneurs to get their ideas onto the political agenda thus depends on their ability to seize windows of opportunity. While the independence of the three streams suggests a large measure of randomness in the agenda-setting process, Howlett states that some factors increase the predictability of the agenda dynamic. Drawing on these factors, he distinguishes between four types of policy windows.[6] First is the discretionary window, which opens at the discretion of identifiable political actors who possess a large influence over the political stream. For example, some policy entrepreneurs could easily sense opportunities when an ideologically committed person is elected as a new head of government. Second, policy entrepreneurs could just as easily take advantage of a spillover window whereby problems might, by association, enter by means of an already open window. Given current knowledge on the environmental impact of farming practices, it may be difficult to renew industrial pollution standards without opening a window for proponents of higher standards in agriculture. Third, institutional practices routinely open windows. For example, when elections approach, window opening may become very predictable. To illustrate, entrepreneurs nowadays know that healthcare is a sensitive issue that politicians cannot ignore when elections approach. Likewise, Canadian entrepreneurs in the agricultural sector know that agricultural ministers meet once a year during the summer and are likely to treat this event as a routine window. Lastly, there remains a less predictable window that Howlett calls the random window, which refers to the random occurrence of disastrous events in the problem stream. The classic example is a plane crash, which may help policy entrepreneurs to get transportation issues onto the political agenda but which, unlike the other windows, cannot be predicted.

If it is portrayed accurately by such theories, agenda setting appears worrisome from the viewpoint of policy-making performance. Agenda setting depends on the capacity of policy entrepreneurs to seize windows of opportunity rather than on the quality of their ideas to address serious problems. Furthermore, the character of these windows of opportunity appears problematic for issues such as agricultural pollution. Random windows, although rare according to Howlett, project the image of a policy process that is highly reactive rather than capable of anticipating problems.[7] A system that needs an environmental disaster to realize that agricultural pollution is a problem deserving the attention of policy makers can hardly be said to be performing at a high level. Discretionary windows are not much more encouraging. If a country needs to wait for the emergence of an influential political actor who, as an individual, shows concern about agricultural pollution, problem anticipation again appears problematic. Agricultural pollution is far from being an issue that sparks passion or even one upon which a committed political actor might be able to build a career.

Spillover windows, on the other hand, might increase the chances for agricultural pollution to be put on the political agenda. However, as the example above illustrates, a window on an issue related to agricultural pollution must already be open, thereby perpetuating the disconnectedness between the severity of agricultural pollution and agenda setting. From a comparative perspective, spillover windows are unlikely to place agricultural pollution on the political agendas of countries where the problem is most severe. Rather, agricultural pollution is more likely to be treated where policy makers have decided to address related problems. In other words, if policy entrepreneurs were to work this window of opportunity, there is no reason to expect agricultural pollution to be more prominent in France than in the United States or Canada, where the problems are not quite so severe.

Lastly, routine windows should be more important for issues that have a longer history or a higher political appeal than agricultural pollution. Howlett treats elections as routine windows, but entrepreneurs with agro-environmental policy ideas are unlikely to see them that way. Agro-environmental ideas are not sufficiently appealing to attract the attention of politicians, even those who are particularly sensitive to the environment. Furthermore, issues with longer histories are more likely to be the object of laws that occasionally need to be renewed, creating routine windows. Issues with longer histories are also more likely to be the object of institutionalized routines, such as annual meetings of ministers or interdepartmental committees. While guaranteeing the more-or-less constant presence of important issues on the political agenda, routine windows rarely offer an entry point on the political agenda for new issues, including agricultural pollution.

In short, academic assessments of agenda setting raise serious doubts that the first condition for performance identified in the previous chapter can be met: namely, that policy makers should not ignore agricultural pollution. The requisite presence of capable policy entrepreneurs and the nature of policy windows define a process incapable of getting agricultural pollution onto the political agenda in a satisfactory way.

Policy Formulation, Networks, and Performance

Assuming that Kingdon is wrong – that is, that the severity of agricultural pollution is sufficiently problematic to get it onto the political agenda and that policy entrepreneurs are not necessary – performance may nevertheless be hindered at the policy-formulation stage. The agenda only guarantees that policy makers do not ignore a given problem; it offers no guarantee that they will have the capacity to formulate an adequate policy and offers even less reassurance that they will implement it.[8] Therefore, policy formulation theories may be discouraging for those who take comfort in the belief that agenda-setting obstacles can be overcome.

In reaction to the behaviourist tendency to ignore the structure of actors' interactions, and also to account for the gradual replacement of hierarchical interactions with horizontal ones, students of policy formulation are increasingly turning to network metaphors.[9] I begin with a discussion of policy network conditions potentially conducive to good policy-making performances in the agro-environmental sector. Some network forms, I argue, present great potential for producing intrusive, comprehensive, and economically viable agro-environmental policies. Unfortunately, recent theoretical developments suggest that networks are unlikely to meet these conditions. Policy formulation difficulties therefore add to agenda-setting obstacles on the way to good agro-environmental performances.

The Composition of Actor Constellations

There are several methods for studying policy networks, but a tendency to distinguish between two types of actors – that is, those who directly participate in policy formulation and those who act as policy advocates – appears as a broadly accepted idea. Coleman and Skogstad speak of an "attentive public" versus a "sub-government" made up of policy participants.[10] Scharpf also distinguishes between "a subset of primary policy actors ... directly and necessarily participating in the making of policy choices and all other actors that may be able to influence the choices of these primary actors."[11] According to Scharpf, the composition of this subset, the actor constellation, has a distinctive impact on policy-making interactions and policy efficacy. The character of the actor constellation and the negotiation setting may contribute to either avoidance or encouragement of lowest-common-denominator and welfare-diminishing policies.[12]

Welfare-enhancing policies can be produced when the actor constellation enters into "problem-solving." Scharpf calls "problem-solving" a negotiation process in which actors direct their attention toward the "joint creation of better projects or objects." He adds that the "power of problem-solving is the power of joint action."[13] Therefore, for such a process to occur, the actor constellation needs to be consistent with two potentially conflicting principles. First, the joining of actors into a common endeavour requires a minimum of cohesion in actors' cognitive orientations. There is a vast literature suggesting that policy networks are constructed on common worldviews – that is, on paradigms or cognitive frames that provide the required cohesion necessary for communication.[14] As Scharpf points out: "Problem-solving is most likely to succeed if the participants are able to engage one another in truth-oriented 'arguing' about the best possible solution and the best way of achieving it."[15] Second, constructive joint action requires enough diversity that actors can each bring complementary resources to the process (e.g., skills, information). Actors with very similar resources are likely to communicate effectively but are unlikely to have much to learn from each other, thereby curtailing the "power of joint action."

The importance of balancing cohesion with diversity is well illustrated in the area of agro-environmental policy. For example, groups representing farmers could rather effectively convince each other that, on the whole, farming practices are not so damaging to the environment and that incremental change is sufficient. In other words, a constellation comprising only farm organizations would not develop comprehensive and intrusive environmental policies for agriculture. If environmental groups were to join this constellation, some agreement on values would surely be necessary; otherwise, in the long term, the constellation would not be viable. At a minimum, environmentalists would have to agree that modern agriculture cannot be eliminated and farmers' representatives would also have to acknowledge the importance of protecting the environment. Assuming that this minimum threshold of agreement is attained, this actor constellation would achieve the power of joint action to a greater extent than one excluding groups representing the environment. Through "truth-oriented arguing," environmental groups may be able to convincingly identify damaging farm practices, while farmers' representatives might sensitize environmentalists to the economic significance of agriculture. Therefore, it is not unimaginable that policies formulated by such an actor constellation would conform to the criteria of intrusiveness, comprehensiveness, and economic viability.

However, problem solving is a rare occurrence since it requires the setting aside of distributive conflicts.[16] To use the above example again, it is difficult to see how a constellation of representatives from the farm sector and environmentalists can avoid discussing distribution of the costs and benefits of the foreseen alternative environmental policies. That is not to say, however, that the process of "distributive bargains" is necessarily conducive to lowest-common-denominator or wealth-diminishing policies. Scharpf argues that wealth-enhancing policies can result from such a process when side payments can be utilized.[17] Agricultural representatives may be reluctant to accept the adoption of intrusive environmental policy instruments, as such instruments would require significant unproductive investments on the part of farmers. If these representatives occupied a veto position in the policy network, they would surely block these instruments' adoption. However, these same representatives may be convinced of the necessity of adopting intrusive instruments if they obtained a guarantee that farmers would receive compensation or a side payment. In addition, the bargaining process resulting from an even distribution of veto powers between the recipients and the payers guarantees that side payments remain consistent with welfare-enhancing objectives: State officials can refuse a payment whose cost outweighs the benefits of intrusiveness. Naturally, this is not as ideal a situation as problem solving, but it nevertheless meets the criteria of high performance in the agro-environmental sector.

It should again be underlined that actor constellations are distinct from epistemic communities. Actor constellations make policies, whereas epistemic communities produce and diffuse policy ideas. Naturally, epistemic community members can also be part of an actor constellation – a situation that can improve policy making, especially when expert consensus exists.[18] Schneider and Ingram, however, have shown that delegation of authority to experts can also involve sizable risks. Scientists, they notably argue, often perceive much complexity in policy problems and therefore demand more studies, occasioning delays.[19] In any case, the participation of epistemic community members in actor constellations by no means constitutes a prerequisite for performance. More often than not, members of epistemic communities remain outside constellations, content to feed constellation actors with policy ideas. Effective policies depend not on the direct participation of the producers of ideas but on the capacity of an actor constellation to draw information from a diversity of sources without endangering the cohesion necessary for negotiations or for establishing a dialogue.

The Structure of Networks

In Scharpf's framework of analysis, actor constellations do not always operate within a network environment and, in fact, can just as easily be observed within hierarchical environments. However, Scharpf acknowledges that policy making is increasingly taking place within horizontal environments: While horizontal interactions do not have to be structured by networks, they often are, especially because of their contribution to the reduction of transaction costs. In a sentence: The durable patterns of interactions that networks establish clarify expectations, thereby reducing the costs associated with policy-making cooperation.[20]

It should be clear from the discussion above that anyone occupying a position as a veto player matters and that veto positions are defined by the structure of political institutions and networks.[21] Durable patterns of interaction may position one or more particular actors from constellations as veto players. As illustrated above, an environmental group may help fill information gaps on agricultural pollution, and regularizing or institutionalizing the relationship between such a group and policy makers may be perceived by the latter as a guaranteed means to remain "enlightened." Such a relationship must be double-sided in order to last. The environmental group will not content itself with only providing information: Sooner or later, the group will come to expect a participatory role in policy making. Information provision itself can serve to influence policy making, but participation is more significant, as it institutes the group as a veto player in the policy network.

Coleman identifies six types of policy networks, each having different implications for veto players. These networks are differentiated along two dimensions: (1) whether civil society actors are included in the actor constellation; and (2) whether power is distributed evenly between state and civil society actors. Corporatist networks exist when civil society actors participate in policy formulation, and they are distinct from state-corporatist networks and clientelist networks, which embody uneven distribution of powers between civil society and state actors. Civil society actors have the capacity to directly participate in policy making when they are sufficiently integrated under umbrella organizations or peak associations. Conversely, when civil society organizations are fragmented, their role tends to be limited to policy advocacy. Networks may then take the form of pressure pluralism – a network distinct from state-directed and issue networks, which are defined by an imbalance in the distribution of powers. Table 3.1 presents these six types of policy networks.

Corporatist networks distribute veto positions most widely because both state and civil society actors can make use of them. Consequently, one should expect corporatist networks to encourage relatively continuous negotiations between these actors. However, similar negotiations are unlikely to occur in state corporatism and clientelism since these networks bestow veto powers on state or civil society actors, but not on both. In pressure pluralist networks, when civil society groups are fragmented and thus limited to policy advocacy, state actors are not so much perceived as veto players as they are seen as "brokers" between coalitions.[22] Brokers, instead of negotiating, tend to consult, manage broad debates with civil society groups, and occasionally arbitrate when a consensus fails to emerge.

Table 3.1

Policy networks

Balance of power/ role of civil society	Policy advocate	Policy participant
Balanced	Pressure pluralism	Corporatism
Favours state actors	State-directed	State corporatism
Favours civil society actors	Issue network	Clientelism

Source: William D. Coleman, "Policy Communities and Policy Networks: Some Issues of Method." Paper prepared for presentation to the System of Government Conference, University of Pittsburgh, 1 November 1997.

Civil society groups, however, are sufficiently strong to expect decisions relatively consistent with the results of these consultation processes. This contrasts with state-directed networks, which permit state officials to make authoritative decisions that go against the wishes of civil society. Issue networks are not as structured as the previous networks and thus tend to resemble anarchic fields.[23]

In addition to the distribution of veto positions, network structures are significant in that they confer different levels of autonomy to state actors or different levels of embeddedness between states and civil society, and these variables have been closely associated with state performance in the comparative public policy literature.[24] Weiss speaks of "governed interdependence" as the relationship structure most likely to equip the state with the "transformative capacity" required in the sector of economic policy to keep a competitive edge in an environment of global market economies.[25] Governed interdependence "refers to a negotiated relationship, in which public and private participants maintain their autonomy, yet which is nevertheless governed by broader goals set and monitored by the state."[26] In summary, governed interdependence requires strong civil society actors who actually participate in policy making and strong state actors who maintain their capacity to exercise leadership. Transformations are unlikely to occur if the state is dispossessed of its autonomy to propose policy alternatives, but those alternatives are unlikely to be efficient if civil society actors are not sufficiently strong. In other words, governed interdependence is most likely to occur – and transformative capacity to be lodged – within corporatist policy networks. In state-directed and in state-corporatist networks, civil society actors are too weak relative to state actors, whereas the situation is reversed in clientelist and issue networks. Civil society and the state are equally strong in pressure pluralist networks, but as a broker the state cannot exercise the leadership required by governed interdependence, and civil society actors are not sufficiently integrated to enter into negotiations on efficient transformations.

The structure of policy networks may thus influence environmental policy-making performance in agriculture in the following manner. In state-directed and state-corporatist networks, in which state actors are insulated, policies may be suspected of being technocratically designed and of underestimating the seriousness of agricultural pollution or the economic importance of agriculture. The difference between these two networks essentially resides in the capacity of government agencies in state-corporatist networks to utilize civil society organizations in order to implement these inefficient policies. The nature of policy decisions in issue networks is more difficult to predict, as the weakness of the structure increases the importance of random circumstances. In clientelist networks, however, policy decisions are highly predictable, especially in the agricultural sector.

In clientelist networks, farmers would certainly be in a position to block the adoption of intrusive and comprehensive command-and-control instruments and may even possess the ability to distort environmental policies so as to subtly increase the level of subsidy already enjoyed in the sector.

As brokers, state actors in pluralist policy networks cannot be expected to exercise much of a leadership role in the transformation of agriculture into a more environmentally sensitive activity. However, this is different from suggesting that policy changes are difficult to achieve in pluralist networks. The literature suggests that policies are much more vulnerable to mood swings or political fashions in pluralist networks.[27] Mood changes and political fashions are instantly reflected in public consultations, thereby inciting state actors, as brokers, to revise where they draw the line among policy alternatives. Therefore, when the mood is on the side of the environment, farmers may suffer from exaggerated intrusiveness, whereas the environment may suffer when the mood favours farmers.

It seems reasonable to hypothesize that intrusive and comprehensive environmental policies sensitive to the economic viability of agriculture are more likely to be designed in corporatist networks. In corporatist networks, state actors should be autonomous enough to identify the problem of agricultural pollution and play a leadership role in the environmental transformation of the farm sector. Nevertheless, agricultural producers and environmentalists, who are closer to the problem, should be well placed to ensure that the transformation is achieved through an adequate dose of intrusiveness, comprehensiveness, and economic sacrifice. Corporatist networks may not always lead to such interdependently governed policy making, but it is the only policy network type providing for the necessary structural conditions.

To summarize, the policy network literature allows policy formulation conditions favourable to policy-making performance to be identified. First, the composition of actor constellations needs to be properly balanced between cohesion and diversity to enable the force of joint action. Second, the structure of the policy networks should be corporatist to distribute veto positions evenly between state and civil society actors. Third, the state has to have the will and the capacity to exercise a leadership role in a transformation process, and civil society groups must have sufficient strength to make major contributions to policy making. Lastly, when distributive issues among veto players cannot be easily set aside, the possibility of providing side payments has to exist. However, policy networks are not static structures insensitive to changes in their environment. Networks change, and the literature offers little hope for policy-making performance since the form of corporatism meeting the above conditions is rarely the direction of such change.

The New Politics of the Welfare State

Below, inspired by the work of Pierson, I show how policy networks can evolve in troublesome directions and how the perspective on policy-making performance just presented can be applied to a specific sector. Pierson argues that welfare state construction has presided over the development of a "new politics."[28] Years of policy making have profoundly altered the nature of political activities, hence the necessity to revise the theoretical understandings that prevail during phases of state building. The historical-institutionalist understanding proposed by Pierson draws on transaction cost economics. Some economists argue that market decisions in which the number of suppliers is high involve high transaction costs. This situation encourages the institutionalization of relationships between otherwise independent economic actors. Institutionalization reduces the uncertainty associated with strategic behaviour and eliminates the costs of looking around for trustworthy partners.[29] When transaction costs are reduced in this way, economic actors will have few incentives to change institutionalized relationships, even when better institutional alternatives become available.

Such an economic reasoning, Pierson argues, applies to politics. Political interactions also involve high transaction costs that actors will seek to reduce. He contends that, just like economic institutions, public policies can help reduce such transaction costs and in so doing reduce their vulnerability.[30] Inspired by North, Pierson shows that policy making involves high fixed costs that create an incentive for policy makers to stand by their decisions for long periods.[31] In addition, policies have learning effects, a process whereby actors accustomed to the requirements of the policies derive increasing returns that encourage resistance to policy change. Over time, policies also create coordination effects since the commitments they inspire encourage collective action among otherwise individualistic actors. In turn, this collective action can only strengthen the policy trajectory. Lastly, policies encourage self-fulfilling expectations by providing targets toward which actors' expectations can be adjusted, further discouraging frequent and sudden policy changes. In short, because they minimize transaction costs over time, public policies have a feedback effect, locking in policy decisions on given policy paths. Pierson argues: "Policies may create incentives that encourage the emergence of elaborate social and economic networks, greatly increasing the cost of adopting once-possible alternatives and inhibiting exit from a current policy path. Individuals make important commitments in response to certain types of government action. These commitments, in turn, may vastly increase the disruption caused by new policies, effectively locking in previous decisions."[32] Added to the short time horizon created by electoral politics, these conditions should discourage any sudden change in policy trajectories.[33]

This sort of historical-institutional analysis implies a change in the political dynamic between the period of welfare state construction, when policy networks were rather thin, and the new politics period, when powerful networks support a large body of policies that concentrate benefits on specific populations. Participating in the construction of the welfare state in the United States, where a culture of collective action had not been encouraged, agricultural policy development was controlled in the 1930s by few policy entrepreneurs. These entrepreneurs from without and from within the state had little difficulty translating their ideas into proper agricultural public policies.[34] However, following a historical-institutionalist perspective, the adoption of these initial policies should have had a significant impact on the composition of the actor constellations and on the development of policy networks. First, the bureaucratic capacity of the state should have been enhanced in order to properly implement the adopted policies. These new state officials would probably have nurtured a commitment to this initial policy, as they garnered an expertise associated with its delivery. Second, the farmers, on whom the policies concentrate the benefits, should have realized that collective action was necessary to protect their interests. As this motive should have incited the strengthening of farm organizations, farmers' commitment to welfare state agricultural policies should also have been strengthened. Third, the policies should have produced a coordination effect whereby agricultural agencies and farm groups, sharing compatible policy commitments, would establish a collaborative relationship. Depending on the degree of fragmentation of farm groups, this relationship should have taken the form of pressure pluralism or corporatism. Fourth, the expectation-adjustment policy effect should have increased the cohesion of the actor constellation.

The purpose of a constellation and network forged by welfare state agricultural policy should therefore be the protection of the initial policy choices rather than transformation. In these policy networks, state actors might have sufficient strength to act autonomously but are unlikely to have enough distance from the initial policy choice – just like the civil society groups who benefit from it – to exercise a leadership role in transformation. While state agricultural agencies and farm groups may develop a relationship of interdependence through a corporatist network, the constellation of actors is likely to be too homogenous to exercise the "power of joint action."[35]

In OECD countries, the problem of agricultural pollution appears within the context of welfare state agricultural policies. Thus, according to Pierson, any understanding of agro-environmental policy development should be consistent with the nature of the new politics of the welfare state. Agro-environmental policies are designed in a context where farm groups and government agencies share a commitment to welfare state policy choices in agriculture, thereby limiting the range of possible trajectories. These policy networks, committed to policies aimed at improving agricultural production, are unlikely to accept any intrusive or comprehensive environmental policies that counter this objective. Environmental policies, in contrast to typical welfare state policies, produce diffused benefits that provide little encouragement to collective action. In summary, policy networks in the agricultural sector, as shaped by the new politics of the welfare state, are unlikely to embody the power of joint action and the transformative capacity required to attain a high level of agro-environmental performance.

Internationalization and Regionalization

It has been argued that the new politics of the welfare state, which is no longer so new, has been displaced today by the internationalization of politics and by regional integration in some places. Some authors suggest that internationalization has severe implications for the composition and structure of the welfare state's policy networks. In fact, internationalization can be understood as "a process (or set of processes) which embodies a transformation in the spatial organization of social relations and transactions – assessed in terms of their extensity, intensity, velocity and impact – generating transcontinental or interregional flows and networks of activity, interaction, and the exercise of power."[36]

Looking at the level of institutionalization of supranational arrangements and at the level of activism of state actors, Coleman and Perl distinguish four "internationalized policy environments": (1) multilevel governance; (2) intergovernmental negotiations; (3) self-regulatory and private regimes; and (4) loose couplings.[37] These four environments appear particularly useful for highlighting distinctions between European and North American countries. The "new politics" of Pierson, it should be recalled, do not allow any anticipation of significant policy differences between Europe and North America. With a high level of institutionalization of supranational arrangements and a high level of state actor activism, the policy environment in the European Union is undoubtedly one of multilevel governance.[38] Conversely, with weak supranational arrangements but a high level of state actor activism, the policy environment of the North American Free Trade Agreement (NAFTA) is one of intergovernmental negotiations. As these two policy environments develop, actor constellations and policy networks are likely to evolve in two different directions. However, neither permits much hope in the way of policy-making performance. Table 3.2 indicates the direction of possible evolution.

As mentioned above, higher cohesion in an actor constellation may enable problem solving and thus be desirable. However, it was also indicated that cohesion should not be allowed to reach a point where actors' resources are no longer complementary since such a situation endangers the "force of joint action." The literature appears to suggest that this is precisely the type of cohesion one might expect from intergovernmentalism.

Analyses of intergovernmentalism, by no mean homogenous, nevertheless tend to highlight processes with a low degree of toleration for diversity of opinion. Intergovernmentalism has largely been understood through the lenses of international relations scholars who adhere to realist assumptions. Among these assumptions, a prominent idea is that international affairs are the domain of state executives advised by a technocratic elite. Suspicious of any involvement of civil society and legislators in international affairs, realist scholars disregard any international actions besides those carried out through intergovernmental forums. It is interesting to note that in demonstrations against globalization, civil society actors are increasingly responding to the patterns of exclusion created by intergovernmentalism. These demonstrations may be launching a process of transformation of the internationalized policy environment, but as long as it remains characterized by intergovernmentalism, civil society and legislators who might have important contributions to make to policy making are doomed to exclusion.

Free trade has been the main subject of North American intergovernmental negotiations for the past two decades. While important debates still exist among scholars concerning the reasons for this orientation,[39] it is rather clear whose opinions these negotiations exclude (if one is to admit they are exclusive) and what kind of cohesion North American intergovernmentalism engenders. In comparison to the treaties of the European Union, the North American Free Trade Agreement contains very few provisions on social or environmental issues. Free trade, it has been argued, reinforces existing patterns of exclusion, as it narrows domestic winning coalitions around the export sectors.[40] Pushed outside the actor constellation by the North American policy-making context, environmental actors are likely to suffer from a limited capacity to argue for intrusive and comprehensive environmental policy instruments for the agricultural sector, or any other sector for that matter.

Table 3.2

Internationalized policy environments and policy networks

Environments/networks	Actor constellation	Policy network
Intergovernmentalism	Higher cohesion	State-directed
Multilevel governance	Lower cohesion	Pressure pluralist

On the other hand, intergovernmentalism may contribute to breaking the dependence of state actors vis-à-vis the beneficiaries of state programs – a situation that again characterizes the new politics of the welfare state. In an influential article, Putnam has presented intergovernmentalism as an opportunity for state executives to increase their autonomy by playing a two-level game.[41] Playing the two-level game consists of manipulating the perception of what is acceptable at one level – the "win-set" – while dealing with the other level. For example, an argument to the effect that a given environmental norm falls outside the win-set of a Republican Congress might earn American executive officials some concessions from international partners. Likewise, executive officials can increasingly count on international negotiations to win concessions from civil society actors. In the agricultural sector, program compressions were presented as a necessary condition to the conclusion of "crucial" trade agreements.[42] In short, intergovernmentalism might increase the autonomy of policy makers vis-à-vis farm groups and, in so doing, transform policy networks into state-directed networks of a particular kind, as the latter mostly involve departments and officials who prefer reductions in public spending and state regulatory activities.

In the current political and economic North American context, this increased autonomy should be targeted at the neoliberal end of eliminating those state interventions believed to be economically harmful. In other words, such state-directed networks, built on internationalization and supported by bureaucratic capacity only in sectors servicing the economy, should serve to "hollow out" the state in all other sectors. Naturally, where environmental protection presents few economic advantages, such a state-directed network presents itself as a powerful hindrance to the development of intrusive and comprehensive environmental policies. While providing state actors with sufficient autonomy to govern, state-directed networks fail to display the level of actor interdependence Weiss presents as a source of policy-making performance. Certainly, such state direction prevents a number of civil society actors who possess important policy information from participating in policy making.

A multilevel governance policy environment, similar to that of the European Union (EU), responds to a different logic. As Risse-Kappen points out, it essentially encourages domestic actors to form or join transnational organizations in order to efficiently participate in supranational policy making.[43] Pushing this reasoning further, Coleman and Perl suggest: "Transnational policy communities will be *less integrated* than national ones, because of larger numbers of actors, more instability in members, lower institutionalization of interactions, and less agreement of basic ideas and values."[44] Debates exist as to whether this constitutes a welcome or unwelcome change in Europe. Kohler-Koch, for example, argues that policy making through transnational networks of actors institutes a novel form of democratic governance that presents several advantages even in terms of decision making.[45] In contrast, Scharpf argues that gains in policy-making capacity attributable to Europeanization are not sufficient to compensate for the losses in domestic policy-making capacity.[46] The desirability of multilevel governance in Europe depends on whether the diversity it engenders brings new perspectives on policy problems or jeopardizes the minimal cohesion necessary for decision making in an actor constellation environment.

Even though the desirability of multilevel governance remains contested, all appear to agree that it destabilizes corporatist policy networks. It has been convincingly argued that European integration erodes the corporatist tradition of several western European countries. First, the transfer of competencies toward the supranational level resulting from the integration process reduces the capacity of member states to participate in corporatist bargains. Corporatism implies a minimum of reciprocity between the state and civil society organizations. When the state is no longer capable of acting reciprocally, civil society organizations cease to have an interest in maintaining the relationship. Second, the transfer of competencies has not been accompanied by a transfer of the minimal bureaucratic capacity necessary for the establishment of corporatist policy networks at the European level. Studies have shown, Streeck argues, that a minimum of bureaucratic capacity is necessary to intervene over the organization of civil society since the creation and maintenance of peak associations often require the intervention of the state. With a bureaucracy smaller than that of several European cities, the European Commission obviously does not possess the capacity to encourage the creation of European associations encompassing enough to render the establishment of corporatist relationships possible with European policy makers. As a consequence, Europeanization would encourage a transformation of corporatist policy networks into pressure pluralist networks.[47]

Depending on its breadth, the diversity that multilevel governance brings into the actor constellation appears more desirable than the exclusion process created by intergovernmentalism. From a policy-making performance point of view, however, the erosion of European corporatism might not be such good news. I explained earlier the advantages that corporatist networks present in terms of governed interdependence. In addition, European corporatism has often been associated with policy-making efficacy, even in the environmental sector.[48] It is after presenting the advantages of the corporatist bargain that Scharpf laments the loss of policy-making capacity engendered by European integration.[49] In short, both intergovernmentalism and multilevel governance, in theory, are unlikely to influence the evolution of North American and European policy networks in a promising direction in terms of policy-making performance.

Conclusion

This chapter does not seek to present – and certainly does not offer – a comprehensive coverage of the theoretical developments in comparative public policy. Rather, the objective was to show that some popular concerns with policy-making performance find an echo in rigorously articulated policy-making theories. The public is concerned that bureaucracies are too big, that interest groups are too powerful, and that globalization constitutes too ominous a force for policy makers to design efficient policies. All these concerns can make theoretical sense. The new politics of Pierson provides a compelling understanding of the power of interest groups and government bureaucracies. Globalization theories are also convincing and certainly predict dire consequences whether they present the phenomenon as one of intergovernmental relations or as one of multilevel governance. Furthermore, these theories often underestimate the difficulty that problems such as agricultural pollution have in finding their way onto the political agenda. In short, these theories can be construed as negative voices confirming the cynicism of citizens. However, the following chapters will attempt to demonstrate that these theories do not withstand empirical testing. The political reality, it will be shown, is brighter than that presented by these theories. Networks, I will argue, more often than not strike the correct balance between cohesion and diversity and are sometimes capable of achieving governed interdependence. Theoretical constructions, therefore, may be misleading when proposing that networks do not deserve legitimacy and trust.

4
France: A Shift from Low- to High-Level Performance

In the early 1960s, France adopted two guidance laws in an effort to support the modernization of its agricultural sector. These laws provided for the consolidation of farms into larger units, the retirement of older farmers with smaller holdings, and legal and financial measures to support interfarmer economic collaboration. In parallel, the Common Agricultural Policy (CAP) was established to provide border protection, price support, and export subsidies, thus encouraging the development of a productive European agriculture, largely to the benefit of France.[1] Briefly stated, France took advantage of the postwar "embedded liberal"[2] compromise to fully participate in the "protected-development" agricultural policy paradigm, perhaps to a greater extent than other countries.[3]

The result of this encouragement to "productivism" was an increasing geographic concentration of agriculture in France, as well as a more intense use of inputs ranging from specialized feed for livestock to chemical fertilizers and pesticides for crops. As outlined in Chapter 1, the consequences for the environment of such changes have not been insignificant.[4]

Despite such a problematic environmental situation, France might be expected to avoid addressing the problem of agricultural pollution with any policy instruments that would seriously constrain farming practices. The reform of the CAP in 1992 and the Blair House Agreement[5] of 1993 marked the beginning of a transition from the protected-developmental paradigm to a market-liberal paradigm, which promised to expose French farmers to market forces and international competition.[6] These circumstances are likely to engender a forceful disapproval of stringent regulations on the part of farmers who fear for their competitiveness. And French farmers are certainly well placed to exclude environmental opinions from the actor constellation. Even though it seems they lost the battle over the 1992 reform of the CAP, farmers in France remain well organized in a peak association, the Fédération nationale des syndicats d'exploitants agricoles (FNSEA).[7] In addition, the interests of the group are strongly entrenched in a corporatist policy network more likely to produce a lock-in effect on productivist policies than adjustments toward agro-environmental policy performance. The corporatist network was put into place to promote the use of the same modern farm practices now blamed for causing deterioration of the environment.[8] In any case, French policy makers appear unlikely to choose to confront farmers due to the weakness of environmental groups and the nonconfrontational approach of the Ministry of the Environment in sectors outside agriculture.[9]

This expectation proved to be right for the 1980s. The agro-environmental policy instruments adopted during that decade were voluntary and did not intrude into agricultural practices. Existing environmental regulations were only marginally applied to farming, and no efforts were made to extend their reach further into the sector. However, the situation changed rather radically in the 1990s when policy makers adopted a package of policy instruments amounting to high agro-environmental performance. The central objective of this chapter is to make sense of this profound change in policy-making efforts toward the environment.

The European Union, I argue, is largely responsible for the change in France since the pressure it produced for the adoption of a serious agro-environmental policy legitimized the French Ministry of the Environment as a participant in the actor constellation. However, this pressure did not automatically translate into policy change, as agricultural corporatism had a mediating effect. As Scharpf suggests, the position of the FNSEA as a veto player in the corporatist arrangement rendered the provision of side payments necessary, and these payments acted as a protection for the economic viability of farming. While not as ideal as problem solving, this distributive bargain appears to enable welfare gains: It corresponds to what I have defined as high policy-making performance in the agro-environmental sector.

This chapter is divided into two sections. In the first section, French policy-making performance is explored further. In the second section, the evolution of both the actor constellation and the policy network is examined as the source of this performance.

Agro-Environmental Performance

I will begin all three empirical chapters with a careful analysis of agro-environmental policy instruments. An agro-environmental policy instrument is any instrument that has among its goals the prevention, reduction, or limitation of environmental damage resulting from agricultural practices.[10] However, I have excluded the registration of pesticides since such an instrument has more to do with the regulation of toxic substances generally than with agricultural practices *per se*. The emphasis in the case of France is on the change between the 1980s and the 1990s. I will therefore produce two analyses, the first looking at the agro-environmental policy instruments in place during the 1980s and the second examining these policy instruments as adopted or amended in the 1990s.

For both periods, I have associated each policy instrument with one of the approaches listed in Table 2.1. Six of the seven approaches were utilized in this exercise. Some of the instruments were *educational* – that is, the goal of the policy instruments was to educate farmers about sustainable agricultural practices. *Financial incentives* or subsidies were also used to encourage the adoption of environmentally adequate practices by farmers. The *regulatory* approach, commonly used in environmental policy making in general, was relied on to provide command-and-control constraints to farm practices. Some *cross-compliance* measures – an approach that obliges farmers to participate in an environmental program in order to be eligible for the general agricultural support programs – were also noted. Quite unique to Europe, the *endogenous* approach, whereby the environment is promoted as an asset helping farmers to improve their economic situation, is also used marginally in France.[11] Lastly, the *reformative* approach, which seeks a redefinition of farming, is apparent in a limited number of policy instruments adopted in the 1990s. Again, the proponents of this latter approach are generally unhappy with the productivist model promoted by the governments of industrialized countries during most of the twentieth century, and they champion models oriented toward enhancing harmony between agriculture and the land.

The educational approach dominated the 1980s, whereas command-and-control regulations and financial incentives became more important in the 1990s. This mix of approaches, which characterizes the 1990s, is an indication that policy makers, preoccupied by the environment, are not sacrificing the economic viability of agriculture. As I argue, however, the reformative approach is gaining importance in the new millennium, moving France closer to problem solving.

More than a change in approaches, what counts in terms of performance is the degree of intrusiveness and comprehensiveness of the policy instruments. Again, comprehensiveness refers to the range of agricultural practices that a policy instrument may cover, while intrusiveness focuses on the degree of change in one or more agricultural practices required by the policy instrument. If a program can subsidize any of the practices, ranging from nutrient management planning to planting hedgerows along fields, it is more comprehensive than a program targeted at only one of these practices. Similarly, a policy that imposes severe restrictions on agricultural waste management, or that strongly incites an overhaul of dominant farming systems, is more intrusive than one that requires a small one-time investment. As Tables 4.1 and 4.2 indicate, changes in comprehensiveness and intrusiveness between the 1980s and the 1990s hint at a profound improvement in agro-environmental policy-making performance.

The 1980s

The Comité d'orientation pour la réduction de la pollution des eaux par les nitrates (CORPEN) was established in 1984 as a response to the 1980 Hénin Report, which identified several agricultural practices as major sources of water pollution. At its creation, the CORPEN was mandated to consult with agricultural interests as well as with interests somewhat associated with agriculture for the purpose of devising voluntary programs, mostly with educational objectives. The CORPEN's mandate also included the preparation of advice for the minister of agriculture and the minister of the environment. In the 1980s the CORPEN was mostly concerned with water pollution arising from nitrates and, as a result, did not cover a comprehensive range of agricultural practices. Moreover, the program did not intrude significantly on generally accepted farming methods. Rather, the idea was to help farmers gradually develop an environmental awareness.

Article 19 was the first agro-environmental policy in what was then known as the European Community (EC). Following a British request, Article 19 was included in a revised structural policy, CEE 797/85, which authorized member states to spend for the purpose of encouraging agro-environmental initiatives in designated areas. In 1987 the EC decided to finance 25 percent of the cost of projects undertaken under Article 19 with money from the guidance section of the European Agricultural Guarantee and Guidance Fund (EAGGF). Article 19 is an incentive measure in that it provides funding for agro-environmental programs.[12] Prior to the 1992

Table 4.1

Agro-environmental policy instruments in France in the 1980s

Policy Instrument	Approach	Comprehensiveness	Intrusiveness
CORPEN	Educational	Low	Low
Article 19	Incentive	Moderate	Low
Extensification	Incentive	Low	Low
Law on Classified Infrastructures[a]	Regulatory	Low	Low
Code rural	Regulatory	Low	Low
Mountain/ seashore laws	Regulatory	Low	Low
Law for the Protection of Nature[b]	Regulatory	Low	Low
Land use planning	Regulatory	Low	Low

a Loi sur les installations classées

b Loi relative à la protection de la nature

CAP reform, member states were devising programs under Article 19 on a voluntary basis. In France, the programs were decided at the local level and could themselves either be simply incentives or else take on a more reformative approach. The first Article 19 programs in France were authorized only in 1989, and according to observers they were far from reformative.[13] Moreover, Article 19 programs in France were only moderately comprehensive in that they did not cover the range of agricultural practices they could have covered and did not generally intrude into farming practices to any great extent.[14]

In an effort to control overproduction and its agricultural budget, the European Community adopted a series of structural measures of extensification between 1988 and 1989. These measures comprised voluntary set-asides for the grain sector, subsidies for the conversion of agricultural land into woodland, and measures encouraging the extensification of beef cattle production.[15] These measures were somewhat targeted and voluntary, and thus not very comprehensive. On the environmental character of the extensification measures, one interviewee said: "Perhaps there was one sentence in the program that gave it an environmental twist, but the overall objective was clearly to reduce production."[16] It should be noted that the extensification measures, along with other budgetary stabilizers, have largely failed to meet this latter objective.[17]

The series of regulations listed in Table 4.1, ranging from the Loi sur les installations classées (Law on Classified Infrastructures) to land use planning, are general policy instruments that contain minor provisions regarding agriculture or the environment. The 1976 Loi sur les installations classées, for example, creates minor constraints for the expansion of large hog operations. All other livestock production projects are submitted to a departmental sanitary regulation that is even less constraining.[18] Article 232-2 of the Code rural includes punishments for farmers who discharge waste into watercourses resulting in the killing of fish.[19] The mountain and seashore laws[20] and the Loi relative à la protection de la nature (Law for the Protection of Nature) of 1976 constrain farming to the extent that they may require the assessment of agricultural projects in certain areas. Lastly, the land use planning process in France provides for the classification of land according to its aesthetic and ecological values, and agricultural practices can be constrained on this basis. However, in 1992 Rainelli noted, with respect to land use planning in France, a "lack of rigour prevailing in the granting of permissions for installations in protected areas, owing to insufficiently stringent ecological analysis when these areas are defined."[21]

In summary, Table 4.1 shows that agro-environmental regulations in France in the 1980s covered only a small range of agricultural practices and were never very intrusive. The European Community, mostly concerned with overproduction and budgetary control, had not yet shown much interest in inciting member states to adopt agro-environmental policies. Thus France mostly relied on the CORPEN in the 1980s to resolve the agro-environmental problems first identified in the 1970s and further documented in 1980 in the Hénin Report. Despite some negotiations, the CORPEN reflects the preference of the Ministry of Agriculture for voluntarism and education over the preference of the water division of the Ministry of the Environment for an approach tending more toward command-and-control regulations.

The 1990s

The situation described above contrasts markedly with that of the 1990s, when new intrusive and comprehensive regulations were finally adopted, existing regulations were amended to become more stringent, some programs began to take on a reformative character, and agricultural policy itself was reviewed. The policy instruments used in the 1990s are listed and categorized in Table 4.2. I will discuss each of these instruments in turn.

The first noticeable change in Table 4.2 is arguably a rather minor one. Reminiscent of the 1980s, the CORPEN adjusted to the 1990s by paying more attention to phosphorous and by developing an expertise on pollution from pesticides, which had simply been excluded from its jurisdiction in the 1980s. The CORPEN also supported Ferti-Mieux, a voluntary program established and managed by farmers through the Association nationale pour le développement agricole (ANDA), whose aim is to provide "guidance" on the evolution of agricultural practices. Lastly, the CORPEN has, somewhat reluctantly, become involved in the implementation of the Nitrate Directive.[22]

The European Community adopted the Nitrate Directive in 1991 as part of its water policy. The directive required member states to identify "vulnerable zones" in their territory, based on a given threshold of nitrogen concentration. At that time, member states were given four years to ensure that no more than 170 kilograms of organic nitrogen per hectare were applied in those vulnerable zones. In addition, member states were required to produce codes of good practice to specify conditions for manure spreading and to make the application of those codes mandatory in vulnerable zones. The Nitrate Directive is particularly important in that it prompted the French government to undertake a series of initiatives in order to meet the limit on nitrogen application set by the directive.

It was partly for this reason that the French government revised its water policy in 1992. The main element of the reform was the definition of a procedure for the elaboration of management plans (Schémas directeurs d'aménagement et de gestion des eaux) for the six watersheds in France. Even more significant for agriculture were the 1993 negotiations to submit agriculture to the water agency system of royalties and aid. The Water Act of 1964 created a system whereby watershed-based agencies were mandated to collect a royalty from those who use or pollute water and to reinvest the money in water quality improvement projects;[23] however, agriculture had been exempted from this system.

In 1992 and 1995, the reach of the Loi sur les installations classées was extended to cover several types of livestock operations as well as smaller operations not previously covered by the law. As a result, a larger number of farms in the 1990s were required either to declare projects in accordance with a general regulation[24] or to go through a process of authorization that included public scrutiny.[25] The Zones d'excédent structurel (ZES) were created in 1993 to impose moratoria on the larger livestock operations in districts *(cantons)* that produce in excess of 170 kilograms of nitrogen per hectare. In addition, the adoption of new technologies and practices to reduce the excess of nutrients was encouraged or made mandatory for certain operations in those zones.

The regulations, ranging from the Nitrate Directive to the Loi sur les installations classées, are moderately comprehensive. They could be more comprehensive in that they are, for the most part, only marginally concerned with crop operations. Nevertheless, they do cover a moderately important range of livestock farming practices. If the resistance of farmers is any indication, it can be argued that these regulations are intrusive.

Table 4.2

New or amended agro-environmental policy instruments in France in the 1990s

Policy Instrument	Approach	Comprehensiveness	Intrusiveness
CORPEN	Educational	Moderate	Low
Nitrate Directive	Regulatory	Moderate	High
The French Water Policy	Regulatory	Moderate	High
Law on Classified Infrastructures*	Regulatory	Moderate	High
ZES	Regulatory	Moderate	High
PMPOA	Incentive	Moderate	High
Landscape Policy	Regulatory	High	Low
PDD	Reformative	High	Low
Set-asides	Cross-compliance	Low	High
Agri-environmental Measures	Incentive	High	Moderate
Rural development/ quality policies	Endogenous	High	Moderate

* Loi sur les installations classées

They impose constraints on farmers with respect to their expansion projects, whereas there were almost no such constraints in the 1980s. In addition, farmers who do not follow more sustainable practices face the risk of paying royalties to the water agencies and other penalties.

The Landscape Policy[26] is not very intrusive into agricultural practices since it only marginally concerns agriculture and, most notably, hedgerows.[27] It is more or less an extension of the seashore and mountain laws to any landscape of a certain aesthetic and ecological value.

Accompanying these various regulations is the Programme de maîtrise des pollutions d'origine agricole (PMPOA), adopted in 1993. The objective of the program is to provide farmers with financial assistance in conforming with the various regulations that I have just described. More specifically, farmers who choose to participate in the program are required to produce an environmental assessment of their farm called a DEXEL *(diagnostic environnemental d'exploitation).* Such assessments are conducted by technicians attached to the Directions départementales de l'agriculture et de la forêt. Following the assessment, a contract on the work to be undertaken is established between the farmer, the water agency, and the state. The cost of the program is shared equally among these three actors. Upon completion of the work, the farmer is no longer considered a polluter and therefore does not have to pay a royalty to a water agency.

A more controversial component of the PMPOA is the schedule whereby farmers become eligible. The PMPOA is a seven-year program that first integrates large farms. For example, in 1994 hog producers raising more than 1,000 pigs were eligible to receive aid under the program, while those raising over 450 pigs had to wait until the year 2001. The budgetary cost of an incentive program is a good indicator of its intrusiveness. In 1993 the government forecast that the program would cost, in total, Fr7.3 billion, among which Fr4.7 billion would be coming from the public purse. A 1999 assessment of the PMPOA suggests that Fr5.5 billion in public funds were spent between 1994 and 1998 and projects that the amount might reach as much as Fr10 billion once the program has reached its objectives.[28]

The Plans de développement durable (PDD), initiated in 1991, are distinguished from the policy instruments discussed above in that they take a reformative approach. Through contracts between the state and individual volunteer farmers, the PDD are aimed at reorienting intensive farming systems toward integration of the production, environmental management, and rural participation functions. In sharp contrast to the logic of intensive farming is the logic of diversification of farming activities adhered to by the PDD. The program is therefore naturally comprehensive, yet it is not very intrusive. In 1996 Fr23.5 million were mobilized for the program,[29] an amount that pales in comparison to the anticipated

Fr1 billion per year for the PMPOA. Since the CAP reform of 1992, the PDD have been one of France's national Agri-environmental Measures.[30]

In 1992 the CAP was reformed to allow agricultural production to be more responsive to market signals.[31] Moreover, set-asides have become mandatory to further reduce production. That is, farmers who do not comply lose their eligibility for the direct payments offered by the European Union (EU) in return for accepting lower prices. It goes without saying that cross-compliance has been strongly opposed by French farmers.

A set of measures called Agri-environmental Measures accompanied the CAP reform. They comprise any measures member states are required to adopt that meet seven objectives, ranging from input reduction to landscape maintenance for leisure.[32] France implemented the Agri-environmental Measures through three national programs and a series of regional and local operations. First was the PDD, discussed above. Second was a program to maintain extensive grassland *(prime à l'herbe)*. The program follows an incentive approach in that it provides payments to farmers who agree not to plough grassland but instead to maintain it, notably by limiting livestock holdings.[33] Similarly, organic farming is encouraged through financial incentives in a third national program. The regional operations are measures defined by regions ranging from input reduction and fertilization planning to biodiversity protection.[34] The local operations, for their part, are basically the continuation of Article 19 on a significantly larger scale, but there are few indications that the local and regional operations are more reformative than previous Article 19 operations. Groups with a reformative agenda might have become more involved in the process of designing such operations, but in the case of conflicts between the various participants, the *préfet de région* provides arbitration, and in the words of an interviewee: "The préfet is always on the side of the strongest, thus never on the side of reformers."[35]

In contrast to Article 19 and the extensification measures of the 1980s, the Agri-environmental Measures are 50 percent financed by the guarantee section of the EAGGF. Thus they are no longer part of the structural policy but are integrated within the main framework of the agricultural policy. This increase in secured resources and in the range of the objectives of the Agri-environmental Measures indicates higher comprehensiveness as well as higher intrusiveness. The anticipated budget of the Agri-environmental Measures for France for the 1993-96 period was Fr4.9 billion, which is still a small amount considering that it represents only about 5 percent of the EAGGF.[36] In addition, the resources are unequally distributed between French programs, with 75 percent of the European Community appropriations allocated to maintaining extensive grassland; therefore, the Agri-environmental Measures are only moderately intrusive.

To these policy instruments, I have added rural development and quality policies in Table 4.2. These instruments are not primarily concerned with the environment, but in the 1990s the environment became a relatively important element of these policies. That is, the environment became increasingly identified as an asset that can contribute to the development of less-favoured areas or that may add to the value of agricultural products. For example, the brand name *"fermier"* indicates products derived from sustainable modes of agricultural production. The European structural and rural development programs, notably Links between Actions for the Development of the Rural Economy (LEADER), feature the preservation of the environment among their main objectives.[37] Out of a 1996 European conference on rural development held in Cork, Ireland, came a declaration that rural development "must be based on an integrated approach, encompassing ... agricultural adjustment and development, economic diversification ..., the management of natural resources, the enhancement of environmental functions, and the promotion of culture, tourism and recreation."[38] In 1995 the EC had approved eighteen programs of rural development in France under LEADER II, worth a contribution of 184,907,000 European Currency Units.[39]

In brief, the agro-environmental policies of the 1980s and 1990s stand in sharp contrast to each other. In the 1980s, education was the dominant environmental approach, whereas command-and-control regulations and financial incentives came to characterize the French approach in the 1990s. France in the 1990s even began experimenting with reformative and endogenous approaches, promising to guide agriculture away from the model valued in that country since the 1950s. As Boisson and Buller suggest: "These various trends point to profound changes in rural France."[40] Specifically, this signified a major policy change from a period during which agricultural pollution was more tolerated and farmers more trusted as environmental stewards to a period during which agricultural pollution became regulated by the state like any other pollution.[41]

More importantly, however, these changes indicate a shift from low-level policy-making performance to high-level policy-making performance. Rather intrusive and moderately comprehensive regulations and programs have been adopted in France since the beginning of the 1990s. These policies stand in sharp contrast to those of the 1980s, when no agro-environmental regulations or programs were intrusive and very few were comprehensive. Furthermore, the provision of financial benefits to farmers, notably through the PMPOA, and efforts to restructure the sector suggest that policy makers have the economic viability of farming in mind when devising environmental policies for the sector.

Recent developments suggest that agro-environmental change had just begun in France during the 1990s. In fact, the government adopted a new guidance law in 1999 to establish a new comprehensive and intrusive instrument, the Contrat territorial d'exploitation (CTE). Largely in line with the PDD, the CTE gradually expands the reformative approach to all Agri-environmental Measures. In fact, the objective of the CTE is to allocate financial assistance to farmers who agree to environmental and social commitments, no longer treating these aspects of farming as inimical to its economic vitality. While the PMPOA benefited large farms first, the CTE attempts to avoid a similar bias, recognizing that small farms, which produce quality food, are just as commercially valuable as larger farms. Therefore, in sharp contrast to the guidance laws of the 1960s, the CTE overshadows the productivist model, trying to align French agriculture with a paradigm of multifunctionality.[42] As an indication of a significant increase in intrusiveness in comparison to the PDD, European Community and French funding for the CTE in 2001 should reach Fr2 billion.

Explaining Performance

The policy network approach presented in Chapter 3 has gained increasing acceptance, particularly in the context of Europeanization, as it efficiently captures aspects of multilevel governance.[43] Marks and his colleagues, along with others, argue that the EU now possesses institutional attributes allowing some horizontal European governance relatively free from the hierarchically constituted national executives.[44] Naturally, several of these institutional attributes are associated with the role and practices of the European Commission, which is a critical actor in policy initiation, an important actor in implementation, a central actor in international negotiations, and a subtle but powerful decision-making actor. I argue that these developments in European governance have engendered important changes in the membership of the domestic agro-environmental actor constellation and in the structure of the domestic policy networks, thereby enabling policy changes that appeared unlikely for France in the early 1980s.

The Actor Constellation

The composition of actor constellations influences the dynamic of policy making and, by extension, policy-making performance. In Chapter 3, I have indeed argued that the actors' cognitive orientations, when balanced between cohesion and diversity, are likely to define a constellation capable of problem solving and thus susceptible to attaining high-level environmental policy-making performance. In short, in order to achieve high-level agro-environmental policy-making performance in the 1990s, the French agro-environmental actor constellation should have changed in composition to accommodate a better balance between the representation of environmental and agricultural interests.

The CORPEN is presented as the result of an agreement between the Ministry of Agriculture and the Ministry of the Environment. However, the educational and nonintrusive approach of the CORPEN suggests that it might be better understood as the result of a bargain or a dialogue focused on the creation of agricultural wealth that largely excluded environmental concerns. In fact, the Ministry of the Environment was rather too poor in resources to advance a strong case for regulation – an approach most notably supported by its water division. Overall, the Ministry of the Environment had little expertise in agriculture at the outset of the 1980s since the ministry was attached to the Ministry of Supply and staffed with *ingénieurs des ponts et chaussées* (civil engineers) when it was created in 1971.[45] The water division of the Ministry of the Environment, as well as the water agencies, had few opportunities to develop an expertise in agriculture prior to the 1990s because the sector was not subjected to the environmental regime set by the Water Act of 1964. In addition, prior to the 1990s, the ministry did not have regional branches and therefore had to rely on the external services of other ministries to enforce its regulations, most notably on the Directions départementales de l'agriculture et la forêt for agricultural matters outside those regarding classified infrastructures.[46] In reference to the CORPEN, an official of the ministry said: "We rarely win when arbitration is necessary; so we have to find solutions that avoid confrontation, otherwise we do not move forward."[47] The Ministry of the Environment in the 1980s, therefore, had little choice but to cooperate with the agriculture ministry on educating farmers about the advantage of using fertilizers "wisely." In short, deprived of the capacity to negotiate or to enter into a problem-solving dialogue, it appears clear that in the 1980s some officials of the Ministry of the Environment chose, albeit reluctantly, to endorse the strategic position of farmers and of the Ministry of Agriculture. Certainly the water division, if not the Ministry of the Environment as a whole, was relegated to the periphery of the actor constellation during this decade and largely fell outside the close circle of veto actors truly engaged in policy formulation.

The strategic position of the water division began to expand within the Ministry of the Environment in the early 1990s as the ministry became strong enough to become a credible policy participant within the actor constellation.[48] The inclusion of the environment in the Single European Act as well as advances in the Community Water Policy at the end of the 1980s have enhanced the position of this division of the Ministry of the Environment in the actor constellation. A former secretary of the environment sympathetic to the strategic position of the water division has stated: "The ministry uses Europe against other French ministries except when the prime minister can accept its point of view. But the environmental battle in France is conducted with the help of Europe ... we always pass by Brussels to win in Paris simply because the environmental point of view is more important for Brussels than for Paris."[49]

In fact, in the early 1990s, the Directorate General for the Environment of the European Commission began to take a more proactive role in the area of water policy by initiating "discharge" directives, including the Nitrate Directive.[50] With a Socialist prime minister rather favourable to an environmental point of view, agricultural interests were at best relegated to the periphery of the actor constellation during the negotiation of the Nitrate Directive. When asked whether agricultural interests were consulted on the Nitrate Directive, one interviewee said: "Absolutely not, the directive was imposed on farmers."[51] Thus the Nitrate Directive was largely negotiated among French and European Community actors (along with actors of other member states) whose views were converging on the regulatory approach. In sharp contrast to the negotiation over the CORPEN, the Nitrate Directive bargain was one of distribution of the costs and benefits of agricultural practices, with little regard for the creation or preservation of agricultural wealth.

However, the negotiation of the French measures required to meet the obligations set by the Nitrate Directive did not take place without farmers. In 1993 the Right returned to power in the National Assembly. Perhaps even more importantly, French farmers were awakened to the necessity of reaffirming their power after the reform of the CAP and Blair House I,[52] both adopted despite their opposition. These steps toward agricultural policy reform appeared outrageous to French farmers, who believed they were victims of these reforms more than any other European farmers.[53] Consequently, farmers from all over France marched massively on Paris in 1993. And in the words of one civil servant occupying important functions within the French state: "Governments in France fear nothing as much as a farmer *jacquerie*."[54] These events helped ensure the inclusion of agricultural interests, most notably those of the FNSEA, in the actor constellation responsible for the formulation of national measures aimed at fulfilling French obligations on the Nitrate Directive. By 1993 the agro-environmental actor constellation comprised a strengthened Ministry of the Environment, free to endorse the views of its water division, and powerful agricultural actors.

The actor constellation of the 1990s appears significantly more diversified than that of the 1980s, thereby enabling the "power of joint action." However, the French Ministry of the Environment, which had displayed significant flexibility on agricultural issues in the 1980s, did not maintain any radical views that trespassed the threshold of cohesion necessary for discussion to occur. But after a decade of environmental experimentation in agriculture, including timid experimentation with the reformative approach, the voice of farmers favourable to environmental innovations appears to have gained importance. The Confédération paysanne, a group supporting alternative agriculture, even obtained almost 28 percent of the votes in the 2001 Agricultural Chambers elections.[55] With such a challenge coming from an environmentally sensitive group, the FNSEA and the Ministry of Agriculture may, in the new millennium, also have an interest in adopting a flexible position. Given the range of ideas that can be discussed in an actor constellation such as this one, it is no surprise that France has adopted the reformative CTE.[56]

Policy Network Structures

Thus far I have demonstrated that the 1980s and the 1990s differed with regard to the composition of the actor constellation. In the 1980s the constellation was composed of actors closely associated with the agricultural sector, whereas in the 1990s it also included actors with environmental concerns. This mix of actors, I argued, created good conditions for high policy-making performance. In Chapter 3, I have underlined that the structure of policy networks also affects the conditions for high policy-making performance. As illustrated below, corporatism distributes veto positions widely between the state and civil society but does not necessarily prevent state actors from playing a leadership role. Such governed interdependence, Weiss argues, should create a transformative capacity likely to result in high policy-making performance.

Since the 1960s the FNSEA has enjoyed a particular corporatist relationship with the Ministry of Agriculture. The FNSEA has important responsibilities for the implementation of the French agricultural policy and, in return, gets to participate in policy making.[57] Fouilleux argues that this specific relationship, known as *cogestion*, has endured even during the CAP reform strongly opposed by the FNSEA.[58] It was under this traditional corporatist arrangement, which excluded environmental actors, that the educational approach of the CORPEN came to prevail in dealing with agro-environmental problems in the 1980s.

However, since the early 1990s, the Ministry of the Environment has become a more central actor in the development of an environmental policy for the agricultural sector, notably because of France's obligations arising out of the Nitrate Directive. Nevertheless, the ministry was initially incapable of ignoring a corporatist tradition in agriculture, by then almost thirty years old. An official of the environment ministry admitted that the "approach in France is based on the recognition that there is a variety of interests ... and accordingly those interests are put around a table to find common solutions instead of allowing unilateral action."[59] Therefore, even though it was characterized by greater diversity in terms of participating actors, the structure of interactions within the actor constellation presiding over the development of an environment policy for agriculture in the 1990s remained corporatist. In other words, the Ministry of the Environment, the Ministry of Agriculture, and the FNSEA occupied veto positions.

Discussions on the integration of agriculture within the system of water agencies began in 1993. Negotiations, as opposed to problem solving, between the veto players proved necessary, as distributional conflicts could not be set aside. More specifically, the FNSEA could not accept that agriculture be regulated like any other sector. One interviewee argued that the FNSEA's part in the discussion was not so much to find the best manner to integrate farmers into the water agency system but to receive financial assistance to help farmers comply with the obligations created by the Nitrate Directive and the 1992 regulation on designated buildings. In his words: "Farmers began making calculations, and they realized that the Nitrate Directive and the new classified infrastructures were going to cost them a lot of money."[60] In contrast, Jean Salmon, the president of the Environment Commission at the FNSEA, claimed that "the water agencies had difficulty thinking outside the rigid scheme that applies to industrial and municipal pollution which has justified their existence since their creation in 1964."[61] This incapacity to set aside distributional issues establishes conditions more conducive to negotiations than to a truth-seeking process.

But the negotiations proved successful, and an agreement was finally reached at the end of 1993 to integrate farmers into the system of water agencies. The FNSEA agreed to do so under specific provisos, notably regarding the schedule and the conditions under which farmers are to be charged a royalty and attributed a substantial subsidy. The PMPOA comprises financial assistance from the water agencies, which also fund other sectors, in addition to important state subsidies. As Scharpf argues, side payments are necessary to prevent a losing player from exercising his or her veto when changes from the status quo are contemplated. And, it should be emphasized, these payments should be viewed as preventing welfare losses rather than as rents.[62] In this case, by preventing farmers from absorbing a significant loss, the payment permitted the project of integrating farming into the water agency system to continue – a situation corresponding to a welfare gain. However, when a welfare gain is environmental quality, measurement is a problem, and thus the gain cannot easily be compared to the cost of the side payment. Nevertheless, it can be assumed that the negotiation setting between farmers and responsible environmental state actors ensures that payments remain roughly within collective utility levels. Specifically, in a state where environmental and financial actors play significant roles, payments such as those enabled by the PMPOA should come to an end when viewed as excessively high in comparison to environmental gains. In line with this logic, the French government has recently mandated a team of financial auditors to assess whether the public expenditures arising out of the PMPOA engender significant environmental gains.[63] The exercise indicates that state actors are concerned about the collective utility of public expenditures, and the veto position of the Ministry of the Environment should guarantee that these expenditures remain within welfare-enhancing levels, although the government still has to act on the recommendations of the report.

In any case, the outcome of the negotiations was satisfactory in terms of policy-making performance, as presented in Chapters 2 and 3. The Ministry of the Environment was able to play a leadership role and deeply transformed the largely unsatisfactory agro-environmental approach of the 1980s by introducing intrusive and moderately comprehensive measures. The close association of farmers' representatives in the process prevented the development of a policy displaying little consideration for the economic viability of agriculture. As the issues of the distribution of the costs and benefits of environmental measures were extensively addressed in the 1990s, it appears now increasingly possible to interact in a problem-solving manner within the French agro-environmental policy network. After almost ten years of proximity within a corporatist policy network, agricultural and environmental state actors sufficiently trust each other to approach problems in a "truth-oriented arguing" rather than a distributive-bargaining style.[64] Arguably, the CTE, a program that promotes a deep transformation of agriculture, would not have been possible without problem-solving interactions.

Conclusion

From the 1980s to the 1990s, France experienced a significant improvement in its agro-environmental policy-making performance. The French environmental approach in the 1980s was voluntary and clearly displayed little inclination to disturb agricultural practices profitable to farmers. This voluntarism disappeared in the 1990s as policy instruments requiring farmers to rely on environmentally responsible agricultural practices were adopted. This, it should be emphasized, did not occur without serious consideration of the economic viability of agriculture. Recent reformative policy developments in fact show that the environment and the economy are increasingly systematically viewed as inseparable.

This change in policy-making performance can be traced to a change in the actor constellation. Although relegated to the periphery of the actor constellation in the 1980s, the Ministry of the Environment became a full participant in agro-environmental policy development in the 1990s. It is interesting to note that the European Union, and the multilevel governance situation it creates, has been largely responsible for this evolution. Allied with European institutions, the Ministry of the Environment obtained the Nitrate Directive, conferring on the ministry important responsibilities in agriculture. In other words, the role of regional integration in the evolution of actor constellations is not as worrisome as some of the theories presented in Chapter 3 have led us to believe.

Likewise, the "new politics" did not prevent state actors – mostly the environment ministry – from playing a leadership role whereby the traditional approach to agro-environmental policy was significantly transformed. Furthermore, farm groups, holding productivist ideas, did not block change but served to sensitize the other participants in the constellation to their economic reality. While the setting aside of distributional conflicts proved difficult in the 1990s, the presence of farmers within the corporatist network effectively complemented the leadership of the Ministry of the Environment, which had to craft policy proposals consistent with the reality of agriculture. To borrow Weiss's concept, governed interdependence occurred, and after ten years of it, there are currently fewer and less salient distributional conflicts, allowing for even greater policy transformations.

5
The United States: Performance in the Absence of Intergovernmental Coordination

It could be argued that the United States is more institutionally predisposed to high-level agro-environmental policy-making performance than France. Serious problems seem more likely to appear on government agendas in the US because the fragmentation of political powers increases the number of locations where policy windows may open.[1] The separation of powers, the weak nature of political parties, and the frequency of elections could affect the stability of political alliances, making a shift of influence possible from agriculturally concerned policy makers to those who are more environmentally concerned.[2] Furthermore, the relationship between the state and private interest groups is not as institutionalized as it is in France – a situation known to allow large variations in the influence of civil society actors over time and over issues,[3] and whereby a transfer of influence from farm groups to environmental groups is possible. Despite these intuitively plausible hypotheses, the US, while attaining a relatively acceptable level of agro-environmental performance, does not measure up to France's performance.

American agro-environmental policy-making performance does not equal that observed in France because coordination has not occurred between initiatives aimed at protecting the environment and those aimed at supporting farmers. Unlike in France, negotiations have not been held in the United States to link the intrusiveness and comprehensiveness of environmental regulations with financial assistance for farmers. Initiatives primarily aimed at protecting the environment in the US have come from state governments, while the federal government's efforts have remained, in a manner consistent with the past, preoccupied with the economic performance of agriculture. This situation carries a lot of potential in terms of agro-environmental policy-making performance but requires a large degree of intergovernmental coordination since aid should naturally be destined to help farmers comply with environmental regulations. This type of coordination, I argue, has not occurred in the US.

Examining the interactions between the membership of the actor constellations and the structure of the policy networks, I have observed that the strength of the United States Department of Agriculture (USDA) prevents the United States Environmental Protection Agency (EPA) from adopting an approach to agriculture similar to that adopted with other industries.[4] The USDA prefers to rely on financial incentives to approach agro-environmental problems. In contrast, state-level water-protection agencies are well positioned to work with state legislatures to develop intrusive command-and-control regulations. However, the USDA, state-level water-protection agencies, and legislatures operate in different and disconnected actor constellations and policy networks; hence it is difficult to direct financial assistance programs to those farmers suddenly targeted by state regulations.

As in the discussion of France in Chapter 4, I will first review the various policies in place at the federal and state levels. I will then discuss the composition of the actor constellation and the structure of the policy networks in order to shed light on the shortcomings of American agro-environmental policy-making performance.

Agro-Environmental Performance

Given the importance of agro-environmental efforts at the subnational level, an analysis of the evolution of agro-environmental policy instruments over the past three decades is more complex for the United States and Canada (the subject of the next chapter) than for France. To overcome this difficulty, I will follow a slightly different analytical procedure for the US and Canada. While all federal agro-environmental policy instruments were included in the analysis, I relied on surveys to identify general trends at the subnational level. In order to obtain a better understanding of these trends, I have selected a limited number of states (North Carolina, Iowa, and Oklahoma) and one province (Ontario – next chapter) for a more careful analysis. This careful analysis is also based on confidential interviews.

The American survey covers sixteen states and clearly points to a general shift toward more state command-and-control regulations. Table 5.1 identifies some of the characteristics of this trend at the state level and lists the federal agro-environmental policy instruments included in the analysis. One notable difference between France and the US is the range of competing approaches to addressing agro-environmental problems. In the US the choice is mainly between approaches that are voluntary in nature, such as the positive-incentive and educational approaches and the command-and-control regulatory approach. The reformative and endogenous approaches have much less institutional support in the US than in France. A convincing explanation for their absence is the historical lack of a peasantry and the hegemony of commercial farming in American history.[5]

A brief look at Table 5.1 already indicates that a number of agro-environmental policy instruments were adopted in the US between the 1970s and 1990s. In consistency with the analysis of France, I will assess each of the federal instruments in turn, focusing on their comprehensiveness and intrusiveness, before turning to the states themselves.

Federal Agro-Environmental Policies

Agro-environmental regulations at the federal level in the US were first experimented with under the Clean Water Act of 1972. Section 502 of the act identifies concentrated animal feeding operations (CAFOs) as producers of point source pollution, thereby subjecting these operations to the National Pollutant Discharge Elimination System (NPDES) permit requirements. In 1976 the EPA adopted a regulation defining what constitutes a CAFO. A CAFO is either: (1) an operation that confines 1,000 animal units for forty-five days or more during a period of one year; or (2) an operation that confines between 301 and 1,000 animal units for forty-five days or more and that discharges waste into US waters. The regulation also stipulates that the permit-issuing authority can designate CAFOs on a case-by-case basis.

However, it should be noted that very few farms confining less than 1,000 animal units are considered CAFOs. Moreover, the EPA estimates that only about one-third of the 6,600 CAFOs in the US have permits and that these are often outdated.[6] Thus very few large farms in the US have a federal NPDES permit.[7] Moreover, manure spreading is considered non-point source pollution and as such is not covered by the Clean Water Act. Therefore, NPDES permits for CAFOs with more than 1,000 animal units may come with a zero discharge rule, but manure applied as a fertilizer is exempted. Therefore, it is fair to argue that the CAFO regulations under the Clean Water Act are neither comprehensive nor intrusive since very few farms are covered and land application is left out.

In February 1998 the Clinton administration released the Clean Water Action Plan. The plan specifically calls for the EPA to publish and implement a new strategy for regulating and permitting animal feeding operations.[8] The strategy was published in March 1998 and promised a revision of the current NPDES regulations for CAFOs, an inspection of all permitted facilities, and a review of the effluent guideline. However, the strategy was quickly subordinated to the *Strategy for Addressing Environmental and Public Health Impacts from Animal Feeding Operations*, a joint USDA-EPA strategy published in March 1999[9] and strangely also a component of the Clean Water Action Plan. Rather accommodating in the protection of farmers' freedom to produce, the unified national strategy suggested prioritizing an updating of the regulations for the largest CAFOs and vulnerable watersheds. Discussions were undertaken and led to the publication of *Proposed Revisions to CAFO Regulations* by the EPA in January 2001.

Table 5.1

Federal policy instruments in the United States and some aspects of state policies

Approach	Federal policies 1970s	Federal policies 1990s	State policies 1970s	State policies 1990s
Regulatory	Clean Water Act	Clean Water Action Plan; House and Senate bills on CAFOs	Few regulations	Permits required; manure management distance planning; local control; separations; production moratoria
Incentive	None	CRP; EQIP; Coastal Zone Act	Limited number of states with a cost-share program	Thirty-one states with a cost-share program
Cross-compliance	None	Deficiency payments linked to Conservation Compliance, Swampbuster, and Sodbuster	None	None
Educational	Land grant universities	Land grant universities; soil conservation service	Land grant universities	Land grant universities

Screening out some of the operations with fewer than 1,000 animal units currently considered CAFOs appears to be their main objective. "Under the three-tier structure, [the] EPA estimates that fewer operations would be designated as CAFOs, with 10 dairy and hog operations maybe designated each year ... [the] EPA has assumed that few animal feeding operations will be designated as CAFOs and subject to the regulation based on historical experience in the NPDES permit program."

The EPA adds that "approximately 3 to 11 percent of all Animal Feeding Operations (AFOs)" would be covered under the CAFO regulations and that the rest "would operate under voluntary programs."[10] Although it is proposed to include the application of manure to land under the CAFO regulations, thereby increasing their comprehensiveness, the granting of discharge permits to encourage the use of treatment technologies is also being considered. The final rules will be known in December 2002, the date contemplated for the adoption of the regulations. Certainty, however, that the subordination of the EPA's strategy to a broader collaboration scheme between the agency and the USDA has curbed the possibility of increasing the intrusiveness of AFO regulations, a trend likely to be accentuated under George W. Bush's administration.

In addition to the administration's plan regarding animal feeding operations, two bills have been introduced by Democratic members of Congress: a Senate Bill in October 1997 and a House Bill in January 1998, both addressing pollution caused by CAFOs. As originally presented, the Senate Bill would apply only to very large operations (over 3,600 pigs). For those operations, nutrient management plans would be mandatory, and land application and the construction of manure storage facilities would be regulated. The bill also calls for strict enforcement of the regulation. The House Bill contains similar provisions except that it would cover a much larger number of operations; it defines a swine operation with more than 1,000 pigs as a CAFO. However, with a strong Republican representation on Capitol Hill since 1994, environmentalists are definitely suffering from a declining influence. As one congressional staffer said: "The Republican majority took care of them ... You can look at a number of votes in the last Congress where the environmentalists were losing ... The bloom is clearly off the environmental rose."[11] In fact, neither of the two bills was passed as Congress currently shows little interest in developing environmental regulations for agriculture. A bill is even currently under study by Congress to exempt some animal feeding operations from permit requirements.[12]

Thanks to the Clean Water Action Plan of the former Clinton administration, federal regulations became slightly more comprehensive in the 1990s (hence a moderate ranking in Table 5.2). As for intrusiveness, however, there is nothing to indicate any significant change, and the election of the Bush administration promises, at best, to preserve the status quo.

Agricultural policy in the US is governed by omnibus farm bills reviewed by Congress roughly every five years, and over the years these farm bills have integrated a number of agro-environmental programs. The 1985 Farm Bill contained four environmental provisions: namely, the Conservation Reserve Program (CRP) and the Conservation Compliance, Swamp-buster, and Sodbuster rules. Conservation programs and services actually have a long history within the USDA.[13] The Conservation Reserve Program was largely inspired by the Soil Bank Program, abandoned by the federal government in the 1960s.[14] The Soil Bank and other earlier conservation measures administered by the USDA were created to retire land from production. Similar motives guided the adoption of the Conservation Reserve Program.

At the beginning of the 1980s, the US had to store important surpluses, mainly because of the emergence of Europe as an exporter of agricultural products.[15] In addition, some US farmers who had made important investments in the 1970s were suddenly badly hurt by the embargo on the USSR following the Afghanistan war.[16] The Conservation Reserve Program was therefore an efficient way to support farmers' incomes without encouraging production.

More specifically, under the CRP a farmer can rent parcels of land vulnerable to erosion to the government for a ten-year period.[17] During this period, the farmer cannot use the rented land for production and must plant it with grass or tree cover. The USDA reimburses half of some conservation expenses on the rented land. In the 1990 Farm Bill, modifications were made to the CRP to enrol land where production was likely to have an impact on water quality.[18] Under the 1996 Farm Bill, the CRP was reauthorized to maintain the level of acreage under conservation contracts, and provisions were made for the development of an Environmental Quality Incentives Program (EQIP), whereby farmers receive assistance to pay for up to 75 percent of the costs of certain conservation practices. Initially, the appropriations for EQIP were in the order of US$200 million annually, but it was increased in the 2002 Farm Bill to reach US$1.3 billion in 2006. The purpose of the increase is to provide for the coverage of a more comprehensive range of practices, including the preparation of nutrient management plans. In the spring of 1998, the secretary of agriculture announced the Conservation Reserve Enhancement Program (CREP), a program that was formally recognized in the 2002 Farm Bill. Although functioning within the CRP framework and budget, the CREP targets specific geographic areas where high-priority environmental concerns were identified. The CREP is also conceived as a joint undertaking between the federal and state governments.

An additional incentive program is Section 319 of the Coastal Zone Act, administered by the EPA. This program provides states within the coastal

Table 5.2

Changes in US federal and state agro-environmental approaches

Approach	Instrument dimension	Federal government		State government	
		1970s	1990s	1970s	1990s
Regulatory	Comprehensiveness	Low	Moderate	Low	High
	Intrusiveness	Low	Low	Low	High
Incentive	Comprehensiveness	Low	High	Low	Moderate
	Intrusiveness	Low	Moderate	Low	Moderate
Cross-compliance	Comprehensiveness	Low	Low	Low	Low
	Intrusiveness	Low	Low	Low	Low
Educational	Comprehensiveness	Low	High	Low	High
	Intrusiveness	Low	Low	Low	Low

zone with funding to implement programs addressing nonpoint source pollution following broad EPA guidelines. The so-called 319 Program has a budget of slightly over US$100 million per year – a small amount in comparison with the more than US$2 billion CRP appropriations for 1994.[19] However, no one should be fooled by these numbers. Since 1985 there have been some ups and downs in CRP appropriations, but overall the levels have been maintained rather than increased. This trend continues with the 2002 Farm Bill. While the CREP appears distinctive, the CRP is in perfect continuation with the Soil Bank Program, whose main objective was to subsidize farmers' incomes without encouraging overproduction. For these reasons, I argue that the federal agro-environmental subsidies became only moderately intrusive in the 1990s compared to previous actions, even though the budget allocated to EQIP in the 2002 Farm Bill announces more intrusiveness. Overall, however, the American federal government mainly responded to severe agricultural pollution by increasing the comprehensiveness of its various programs of positive incentives.

Conservation Compliance, Swampbuster, and Sodbuster rules are designed to penalize farmers whose production practices might endanger soil or wetlands. Under these rules, farm program benefits are cut off for producers who do not carry out an approved conservation plan on land vulnerable to erosion (Conservation Compliance) or who farm either on wetland (Swampbuster) or on highly erodible land (Sodbuster). However, in the 1990 Farm Bill, the penalties under the Swampbuster rule were significantly reduced, while the Sodbuster and Conservation Compliance rules were merged. Observers have complained about the lack of enforcement of the Conservation Compliance provision,[20] which was incidentally made even less intrusive in the 1996 Farm Bill. The 2002 Farm Bill largely leaves these programs intact.

Finally, agricultural education and extension in the US are the joint responsibilities of the federal and state governments, and the land grant universities and conservation districts, in which agricultural pollution is a concern, deliver the programs. In fact, conservation districts have been providing important extension services to farmers in collaboration with state and federal conservation services for a long time. Among other things, they provide engineering guidance for the construction of manure storage facilities. It is obvious that environmental concerns have gained importance for land grant institutions over the years.[21] But, unlike in 1980s France, the educational approach has never been promoted in the US as the main approach to agro-environmental problems.

It appears clear from this brief analysis and from Table 5.2 that between the 1970s and 1990s, the US federal government refrained from adopting intrusive and comprehensive regulations. Even the regulatory projects under study at the time of writing promise to exclude a large number of farms from important restrictions. Instead, the US government opted for reviving and revamping conservation programs with the result that in the 1990s, in contrast to the 1970s, the government had rather comprehensive incentive programs. When compared with conservation programs the country has had since the 1930s, however, the change is marginal as far as the environment is concerned since the primary objective of the CRP was to subsidize farmers during a difficult period. Environmental objectives became more genuine only when EQIP was adopted and when a decent budget was allocated to the program in the 2002 Farm Bill. Nevertheless, with few command-and-control regulations, and with incentive programs characterized by distinct official and unofficial environmental objectives, a focus on the federal government necessarily leads to a conclusion of low agro-environmental policy-making performance. However, when the states enter into the analysis, the portrait changes.

State Agro-Environmental Policies

As indicated above, the task of reviewing and analyzing the regulations and programs of the entire fifty states would be overwhelming. However, two researchers from Purdue University, Alan Sutton and Don Jones, have recently conducted a survey of animal waste management regulations in sixteen states that reveals some trends worth mentioning.[22]

Comparing their survey to one conducted in 1992, the two authors conclude: "It is clear that individual state regulations are currently more stringent in storage structure design and approval, more separation distances implemented, more attention and requirements on land application rates, and development of manure management plans ... [And] a most significant change in regulations currently is the attempt to control odor problems with setback distances."[23]

Of course, the authors note variations in terms of rigour in some other aspects of state policies, notably in lagoon closures, in the importance of separation distances, and in infiltration area requirements. But what is important to note here is the trend in most agricultural states toward the adoption of CAFO legislation dealing with a rather wide range of practices. My own analysis of three states confirms the trend but not without adding significant nuances.

To bring out these nuances, I selected three somewhat dissimilar agricultural states. The first is Iowa, which has the largest swine inventory in the United States and whose structure of production remains largely characterized by family farming. The second, North Carolina, offers an interesting contrast to Iowa in that a narrow network of corporate farms pioneered the substantial and recent growth of the livestock industry. Lastly, Oklahoma is one of the few Midwestern states that has repealed its law restricting corporate ownership in recent years, thereby opening its doors to large integrators. In short, Iowa has traditionally had a reputation for being a friendly place for family farms, and North Carolina has enjoyed a reputation for providing a good climate for corporate farms, while Oklahoma is attempting to attract larger corporate farms.

Despite these differences, it is interesting to note that each of the three states adopted laws in the 1990s penalizing livestock farmers whose practices might pollute, thereby confirming the trend toward more regulations. However, the statutes of the three states are not equally stringent for every farmer. The regulations are compared on a number of aspects in Table 5.3.

North Carolina has the most stringent regulations since all farms of more than 100 animal units were required to have a permit issued by the Department of the Environment, Health, and Natural Resources before December 1997.[24] The permit process comprised the certification of animal waste management plans covering waste collection, storage, treatment, and application. Moreover, permitted facilities are to be inspected frequently to ensure that standards of operation and maintenance are respected. In the face of inspection difficulties, the state has increased the staff of the Division of Water Quality, the state agency in charge of carrying out this task. Moreover, statutes passed between 1995 and 1997 require that operators who apply manure on land be certified, impose separation distances for new buildings and manure application on land, implement county zoning, and impose a two-year moratorium on the expansion of the hog industry.[25]

The permit requirement in Iowa covers fewer farmers. Coverage varies, but farms with fewer than 400 cattle or 3,600 pigs are exempted, as are larger farms that use formed-manure storage structures or solid-waste systems. Moreover, permits are required only for new and expanding operations. As in North Carolina, permitted operations in Iowa must have manure management plans. Rather severe minimum distance separations must also be respected by farmers. And, unlike in North Carolina, the new facilities that require a permit in Iowa have to meet certain environmentally oriented construction standards. Lastly, reflecting Iowa's accommodating nature toward farmers, the state prohibits county regulations and has not seriously considered imposing a moratorium on the expansion of any livestock industry.[26]

In terms of regulations, Oklahoma sits between Iowa and North Carolina. On the one hand, the state has imposed severe regulations similar to North Carolina's, notably a moratorium on the hog and poultry industries. The state has also authorized local control, imposed constraining distance separations as well as construction standards for lagoons, and called for inspections and important fines in case of violation of the law; it has even required farmers to adopt odour abatement plans. On the other

Table 5.3

Livestock regulations in three US states

Measures	Iowa	North Carolina	Oklahoma
Permit requirement	Less encompassing than the federal NPDES*	More encompassing than the federal NPDES	Less encompassing than the federal NPDES
Manure management plans	Required for permitted operations	Required for operations of 100 animal units or 250 pigs	Best management and odour abatement plans required for permitted operations
Separation distances	New buildings: 1,000 feet; Land application: 750 feet	New buildings: 1,500 feet or more; Land application: 50 feet	New buildings: ¼ of a mile or more; Land application: 500 feet
Local control	Disabled	Enabled	Enabled
Moratorium	No	Yes	Yes

* The federal NPDES permits are mostly for operations with more than 1,000 animal units, the equivalent of 2,500 pigs and 1,000 bovine. There are some small differences between the results reported in this table and those of Sutton and Jones.

hand, the requirement to have a permit applies only to operations of 2,000 animal units (the equivalent of 5,000 pigs), which is twice the threshold of the NPDES federal permits. Moreover, the regulatory authority, unlike in the other two states, is lodged within the agricultural rather than the environmental agency.[27] In other words, Oklahoma is tough but, as a general rule, only with very large farms.

In short, as Table 5.2 suggests, states where agriculture is an important activity adopted regulations in the 1990s that are more comprehensive and certainly more intrusive than the federal regulations. Table 5.3 further indicates that from one state to the next, regulations impose constraints on farming practices that vary slightly in substance and rigour. I will now turn to the provision of agro-environmental financial incentives by state governments.

According to a recent report, thirty-one states have established conservation cost-share programs.[28] In general terms, cost-share programs are funded strictly with state money and are designed to match a given percentage of farmers' conservation investments. Of the three states under study here, Iowa has the oldest program. It was established in 1973 and now provides farmers with up to 60 percent of the cost of certain conservation practices. In 1997 the budget of the program was US$6.6 million and it has cost a total of US$144 million since 1973. In North Carolina the shared cost program was established in 1984 and covers "up to seventy-five percent of the cost of practices designed to protect soil and water, including improved animal waste management."[29] In 1995 funding for the shared cost program had reached US$8.2 million. And in 1996 House Bill 53 added US$5.75 million to the program, in addition to reserving 6.5 percent of any remaining balance from the General Fund for the purchase of land, including agricultural land, for the purpose of creating riparian buffers.[30] However, the shared cost programs in Iowa and North Carolina are small programs when compared to the French, American, and even some Canadian incentive programs. The programs are nevertheless rather generous compared to similar programs in other states. Oklahoma, for example, has yet to adopt such a program.

None of the states has used cross-compliance to promote the use of conservation practices. This is not very surprising given that the states, unlike the federal government, have few general agricultural programs supporting farmers' incomes. Again, education and extension are provided by land grant universities and conservation districts, and both are federal-state cooperative endeavours.

In summary, the federal government has very few regulations concerning agricultural pollution, and at this point the evidence suggests that this situation will prevail in the near future. Most federal actions in the area of agro-environmental policy have taken the form of reviving and revamping old conservation programs that, in fact, date back to the 1930s. From the 1970s to the 1990s, federal incentive programs nevertheless became more comprehensive, covering a wider array of farming practices. With the 2002 Farm Bill, they may even be made more intrusive. As Table 5.2 suggests, the situation is almost the opposite at the state level. The incentive approach has remained a secondary instrument for states, while regulations dealing with agricultural pollution have become comprehensive and intrusive, with some variation from one state to another.

When the United States is considered as a whole, agro-environmental policy-making performance might appear well served. In a manner comparable to France, the US has a rather appealing mix of intrusive command-and-control regulations as well as financial aid programs. However, in contrast to France, the financial programs are not conceived as a compensation for intrusive state regulations, and therefore most of the money channelled through these programs fails to help farmers comply with these regulations. The CRP, by far the most generous of these financial programs, is geared toward land retirement and mainly aims to reduce surplus crop production, rather than to help farmers prepare required nutrient management plans, abide by mandatory construction standards and distance separations, or survive a moratorium on production growth. As a collaborative effort between the federal and state governments, the CREP is an improvement, but it remains a marginal part of the CRP. EQIP promises important environmental improvements, but as an initiative that excludes intergovernmental coordination, it offers no guarantee that the norms it embodies are harmonized with the state regulations to which farmers need to respond to avoid sanctions. In short, while significant agro-environmental efforts are being undertaken in the United States, the performance is not quite as satisfactory as that observed in France in the 1990s.

Explaining the American Performance

Because agro-environmental policy decisions at the state and federal levels appear rather disconnected, it seems appropriate to study the composition of actor constellations and the structure of policy networks at the state and federal levels separately.

The Federal Constellation and Network

An important difference between France and the US is that in the latter country, interest groups are rarely organized into peak associations such as the FNSEA.[31] The fragmentation characterizing the American interest group system makes it difficult for such groups to participate directly in policy making. In other words, negotiated agreements between state agencies and civil society groups, such as those observed in France, can only be exceptional in the American institutional context.

This variation in the situation of groups, I argue, is likely to affect their policy preferences. One might expect groups in a fragmented system to have policy preferences more adaptable to new circumstances. For example, farm groups in the United States, aware of their limited individual power potential, might be more willing than their French counterparts to accept that they are responsible to some extent for the degradation of the environment. The work of Sabatier provides useful insights for studying the evolution of the preferences and beliefs of policy-making actors.[32] For Sabatier, policy change is often the result of two somewhat interactive processes. First, external shocks create perturbations within a policy subsystem, with an effect similar to the opening of Kingdon's policy window.[33] Second is policy-oriented learning, which is defined as the "alterations of thought or behavioral intentions ... which are concerned with the attainment or revision of the precepts of the belief system of individuals or of collectivities."[34] According to Jenkins-Smith and Sabatier, learning takes place "within" or "across" advocacy coalitions[35] – that is, it involves communicative action among the members of a coalition or between the members of two or more coalitions. Major policy changes are likely to occur when learning takes place between advocacy coalitions, whereas learning within advocacy coalitions leads to more incremental changes.

The power of the FNSEA, derived from its quasi-monopoly situation in terms of French farmer representation, provides no incentives for the organization to engage in a learning process with groups holding different views on agriculture, such as environmental groups. The FNSEA simply does not need to enter into a multigroup coalition to efficiently exercise policy influence. Conversely, in a system of interest group fragmentation, where governments tend not to collaborate directly with encompassing organizations on a continuous basis, the exercise of influence is much more dependent on the formation of multigroup coalitions or, in the language of Sabatier and Jenkins-Smith, of advocacy coalitions. As these two political scientists point out, the formation of such coalitions requires communicative action and possibly a large dose of learning – that is, the alteration of beliefs and policy preferences.

Group activities in the agro-environmental sector in the United States support this proposal. Public pressure for regulating the hog industry increased in the US following changes in the structure of production and giant manure spills, notably in North Carolina. In reaction, the National Pork Producers Council (NPPC) increasingly sought discussions and support from several groups on appropriate solutions. According to one interviewee, the council "spent a lot of time in Washington asking whether there was a way to get different people to come together and talk about these issues."[36] Policy learning and the process of forming an advocacy coalition clearly began when America's Clean Water Foundation, a group of state water quality regulators, accepted to undertake a "Pork Dialogue."[37] Environmental groups, hog producers from a number of states, the EPA, and the USDA were invited to participate. Large environmental groups were the only invited parties that declined, but not without engaging in an informal dialogue.[38] During the process, the formal participants in the Pork Dialogue consulted with various experts, other farm groups, and the public.

As a result of the dialogue, the NPPC came to endorse the position that it was necessary to regulate hog production. While the regulatory framework proposed by the dialogue excludes regulations such as production moratoria, it includes several proposals for regulations that are sometimes more severe than those in place in the three states discussed above. For example, the dialogue "calls for review and approval, by the appropriate regulatory authority, of all pork production operations, *regardless of size*."[39] In addition, the dialogue suggests construction standards for facilities, certification of the operators, important setbacks for the construction of farm buildings, and the land application of manure, as well as severe penalties for violators.

There is no doubt that the endorsement of such proposals by the National Pork Producers Council came as a result of learning the potential benefits of regulations for hog producers. About the Pork Dialogue, one member of the council said:

> EPA and state regulators presented data that small operators were in fact chronic polluters in parts of the country. Then someone said that bigger operators, because they have permits, have the ability to demonstrate in a court of law that they are not polluting, but all those small operations that are exempted from the regulation do not have that same ability to protect themselves. Anyway, it is not right to say to a group of people that it is OK to pollute ... [Under those circumstances,] we came to think that farmers can afford to document that they are not spreading their manure over the top of wells or near surface water for example. And that notion caught on.[40]

Most general farm organizations, such as the American Farm Bureau, were undecided at the end of the Pork Dialogue as to whether to fully endorse its regulatory approach, but it is reasonable to argue that their position has also evolved as a result of the process. As one USDA official said, the result of the Pork Dialogue "is a heroic document." Given the importance of learning, then, it may be surprising that the actors within the federal actor constellation did not embrace the command-and-control regulatory approach.

I argue that advocacy coalitions are more likely to be successful at influencing the preferences of the actors within a constellation, and thereby at inducing policy change, in pressure pluralist policy networks than in any other type of policy networks. And I further suggest that with the USDA imposing itself as a strong bureaucracy capable of leadership and not just brokerage, the federal agro-environmental policy network corresponds to state direction rather than to pressure pluralism (see Table 3.1). Over the years, the USDA has favoured an understanding of agricultural pollution centred around the notion of preserving agricultural wealth, and pressure to integrate the idea that farmers must change their practices to meet broader environmental norms has been resisted. As the leading agency in a state-directed policy network, the USDA was able to translate this preference into policy, ignoring any learning arising out of the Pork Dialogue.

As discussed in the previous section, environmental concerns at the US federal level have been largely addressed in farm bills, which are naturally oriented toward the generation of wealth for farmers. Traditionally, these bills have been of interest to members of Congress from agricultural states or districts. These members are well represented on the agricultural committees of both houses, and more importantly they are supported by the USDA, the agency responsible for the implementation of farm bills.[41] These mutually supportive actors were central when the views of environmentalists gained acceptance during the negotiations of the 1985 Farm Bill endorsing the provision of agro-environmental financial incentives. Not so surprisingly, this policy approach was in line with the USDA's long experience with conservation policy.[42]

Coinciding with rising public pressure for agro-environmental regulations in the 1990s, the former Clinton administration mandated Al Gore, then vice-president, to lead an important initiative: the Clean Water Action Plan. With environmental issues off the agenda of the Republican Congress, the Clean Water Action Plan sought to move on regulating agricultural practices without calling for legislative actions. As one interviewee said: "The Clean Water Action Plan represents a revision of the Clean Water Act without having to go through the Republican Congress."[43] Another interviewee pointed out that the Clean Water Act allows some discretion for the administration and the EPA to act on the CAFO issue without Congress.[44]

However, in the execution of the plan, the EPA had to confront the USDA, which possesses much more experience in dealing with agro-environmental problems. The Natural Resources Conservation Service (NRCS) of the USDA has a history dating back to the 1930s. Moreover, the NRCS is a rather resourceful service with access to thousands of state conservation districts that have a lot of experience in implementing conservation policies.[45] In contrast, the EPA has a short history and a relatively small field-level staff spread over ten regions, ill-coordinated by a central office.[46] Generally speaking, the USDA has for a long time been a powerful federal department in comparison to the EPA. In fact, few bureaucratic agencies of the American federal government possess the strength of the USDA.[47] It is in this context, whereby the USDA can exercise state direction, that one must understand the failure of the Clean Water Action Plan to produce agro-environmental command-and-control regulations. As one USDA official said: "We have a policy position of no pollution at large ..., but the process of getting there is through technical assistance, conservation planning, and cost-shared incentives."[48]

State Constellations and Networks

The Pork Dialogue, and environmental pressure in general, has had more of an impact on state actor constellations, since state agro-environmental policy networks take the form of pressure pluralism. To a large extent, the centralization of agricultural policy making around the USDA – a particularly strong bureaucratic agency – has pre-empted the development of similarly powerful agencies at the state level. As a result, state departments of agriculture are minor players in agricultural policy making in general and in agro-environmental policy making in particular. In fact, responsibilities for conservation policy at the state level are often located in departments of the environment and natural resources, rather than in departments of agriculture. In short, in contrast to the USDA, state departments of agriculture are relatively weak and lack a policy tradition in the agro-environmental sector, making them incapable of state direction.

In such a context, state environmental agencies and state legislatures have been free to respond to pressure favourable to intrusive command-and-control agro-environmental regulations or to broker in favour of groups supportive of more traditional environmental policy instruments. Some states have a history as pioneers in a number of issues relating to environmental protection[49] and were somewhat supported in their endeavours to tackle agricultural pollution by the EPA. Although modest in budgetary terms, the Section 319 Program – a grant program administered by the EPA since the end of the 1980s – helped states turn their attention to agricultural pollution. A North Carolinian agricultural official explains that the state capacity to enforce severe agro-environmental regulations can be linked to Section 319: "The federal-funded programs for municipal waste were rapidly decreasing and the Department of the Environment and Natural Resources had a lot of employees working on this. At the same time, [the] EPA began providing money for nonpoint source pollution. And it is a fact that a lot of the people working for the municipal program moved to animal waste."[50]

In short, new environmental concerns entered quite different constellations at the state level than at the federal level. The state constellations comprised actors displaying a relative capacity to enforce agro-environmental command-and-control regulations and did not comprise actors normatively predisposed and capable of resisting the adoption of such policy instruments. In contrast to the federal constellation, state actors at the subnational level saw their role as brokers in pressure pluralist networks rather than as leaders in state-directed networks.

It should be clear at this point that the lack of agro-environmental coordination identified above is attributable to the fact that the federal state-directed network and the subnational pressure pluralist networks are largely disconnected. While aware of the general situation at the state level, the USDA has shown little inclination to establish cooperative relationships with subnational environmental officials. Typical of state direction, federal agro-environmental programs are developed in a technocratic fashion and are ill adapted to the variable regulatory context of individual states. In other words, federal programs do little to help farmers comply with intrusive state regulations, which is alarming since American subnational pressure pluralism, in comparison to French corporatism, enabled sudden policy shifts that left little time for farmers to prepare to change their agricultural practices.

Conclusion

If the American performance does not appear to be as ideal as that observed in France, it should be underlined that it still remains far from the doom scenarios discussed in the earlier, more theoretical chapters of this book. First, as a whole, the United States has an interesting mix of command-and-control and financial aid agro-environmental policy instruments. Second, the absence of intergovernmental coordination, or the divergence between the federal and state agro-environmental approaches, has not resulted in the adoption of policies with conflicting objectives. The federal government agro-environmental effort does not cancel out the efforts of state governments. Arguably, the absence of coordination has minor economic consequences.

As for interest groups, they did not conform to the negative image drawn by the new politics of the welfare state theory. The American situation, and particularly the Pork Dialogue, shows that interest groups not only are involved in policy making to protect the level of benefits they enjoy from existing policies, but are capable of learning – that is, of adapting their policy discourse to the management requirements of emerging problems. In fact, this chapter shows that performance is not so much affected by interest groups themselves as it is by the structure of policy networks. The USDA failed to adapt its policy preferences at the pace of interest groups because the agency is capable of acting as a leader in a state-directed policy network. The absence of coordination, the main agro-environmental policy problem in the US, is due to the presence of largely disconnected actor constellations and networks, a common problem in federal systems, in this case accentuated by the USDA's state direction. Interestingly enough, despite the significant coordination problems it often creates, federalism is rarely a source of pessimism among citizens, especially in the United States.

6
Canada: Stalled at a Low Performance Level

Canada is internationally known for its large open spaces and unspoiled natural environment. The country generally enjoys a good environmental reputation. However, as the title of this chapter indicates, Canada's agro-environmental policy-making performance does not compare favourably with the performances of France or the United States. A large proportion of Canadian open space is in fact under cultivation, and policy efforts devoted to protecting the environment in agricultural areas appear to be insufficient. This is surprising since Canadian governments have committed themselves to protecting the rural environment. In 1995 the federal and provincial ministers of agriculture promised to coordinate their efforts "to effectively deal with the challenges and seize the opportunities presented by increasing public interest in environmental sustainability."[1]

Canada is a federal country like the United States, albeit even more decentralized. As such, Canada could also be expected to suffer from an absence of positive intergovernmental coordination.[2] While intergovernmental coordination is a potential source of problems for Canada,[3] the source of the real problem is located within the federal and provincial actor constellations. As this chapter will demonstrate, the federal actor constellation shows little inclination and capacity for adopting intrusive and comprehensive policy instruments. Since the 1980s, when agricultural pollution landed on the agenda of the federal government, policy efforts have been limited to gathering information on agricultural pollution. On the other hand, farmers, through clientelist networks, often control policy development at the provincial level and thus remove comprehensive and intrusive policy instruments from the government arsenal. In fact, as we will see, the regulation of agricultural practices in Ontario is considered not so much a responsibility of the state as "farmers' business." In short, in contrast to France and the United States, Canada possesses a limited capacity for dealing with agricultural pollution and displays little inclination to tackle the problem.

This chapter begins with a presentation of the policy instruments in place at the federal and provincial levels, focusing particularly on Ontario. Then, as in the previous two chapters, I will analyze the actor constellations and policy networks in greater detail.

Agro-Environmental Policy Divergence

This final empirical analysis begins with a review of all the federal agro-environmental policy instruments, and after a survey of the general situation at the provincial level, it concludes with a review of Ontario's agro-environmental policy instruments. Ontario has been selected because the character of agro-environmental policy making in that province is fairly close to that found in most other Canadian provinces. Moreover, Ontario provides an illustration of the functioning of a policy network type – clientelism – not encountered in either the United States or France. Table 6.1 lists all the federal and Ontario policy instruments already in place or developed during the 1990s.

The Federal Agro-Environmental Effort

The Canadian federal government, despite its environmental commitment,[4] has chosen to refrain from adopting any intrusive agro-environmental programs, including financial incentive programs, and has especially avoided command-and-control regulations. Table 6.1 lists the Canadian Environmental Protection Act (CEPA), an act revamped in 1999, as a command-and-control policy instrument. However, the CEPA has very little to do with agriculture. Sections 116 to 119 of the act do address nutrients, but they restrict the regulatory power of Environment Canada to cleaning products. The rationale invoked for exempting agricultural fertilizers was that the Fertilizers Act already regulates them.[5] However, the Fertilizers Act deals with the marketing of fertilizers and the setting of labelling and toxicity standards; it does not address their utilization by farmers.[6] This act obviously fails to regulate the land application of unprocessed agricultural wastes.

The CEPA does offer moral support for the National Program of Action for the Protection of the Marine Environment from Land-Based Activities (NPA), whose objective is to identify priorities in terms of pollution abatement on land and to recommend improved planning and management processes where appropriate. However, the program gives agricultural pollution a low priority, with higher priority placed on municipal and industrial sources of pollution.[7]

The NPA reaffirms the importance of ecosystem initiatives such as the St. Lawrence Action Plan Vision 2000, a program whose current emphasis is agricultural pollution. Yet the approaches favoured by the St. Lawrence Action Plan and other ecosystem initiatives sponsored by the federal

Table 6.1

Canada's federal and Ontario agro-environmental policy instruments

Approach	Federal	Ontario
Regulatory	Canadian Environmental Protection Act; Fertilizers Act	Environmental Protection Act; Ontario Planning Act; Farm Practices Protection Act; Ontario Water Resources Act; Conservation Authorities Act
Cross-compliance	N/A	Certificate of compliance is required by some municipalities to obtain building permits
Incentive	Agricultural Environmental Stewardship Initiative (AESI)	Financial incentive for the implementation of an Environmental Farm Plan (EFP) program
Educational	National Program of Action for the Protection of the Marine Environment from Land-Based Activities (NPA); Ecosystem Initiatives Agri-Environmental Indicator Project	Best Management Practices booklets; Watershed Management Policy
Self-regulatory	N/A	Environmental Farm Plan (EFP) program

government do not rely on command-and-control regulations or on the provision of grants in exchange for the adoption of low environmental impact agricultural practices, as in France and the United States. Instead, the federal government generally prefers to rely on education or moral suasion.[8]

The Agri-Environmental Indicator Project is arguably the most important agro-environmental initiative of the federal government. Far from being a regulatory initiative, the project's objective is to produce scientific information on the environmental impact of agricultural practices. This scientific information, it is assumed, will demonstrate "the progress being made by the agriculture sector" and support "the development of strategies and actions targeted at areas and resources that remain at environmental risk."[9] Interestingly, the state agency responsible for the Agri-Environmental Indicator Project is not Environment Canada but Agriculture and Agri-Food Canada. Accustomed to providing services to farmers, Agriculture and Agri-Food Canada prefers educating farmers about the environmental impact of their practices rather than enforcing environmental regulations.

Lastly, the Canadian Adaptation and Rural Development (CARD) fund, with Cdn$60 million per year, is aimed at helping "the agriculture and agri-food industry adapt and grow to meet the challenges of a rapidly changing world economy."[10] CARD cites the acceleration of sustainable environmental practices among its six specific objectives. To meet this objective, the Canadian minister of agriculture recently announced the Agricultural Environmental Stewardship Initiative (AESI), a Cdn$10 million initiative that should encourage the province-based and industry-led Adaptation Councils to spend more of their funds on environmental projects. The AESI is classified in Table 6.1 as a financial incentive, but unlike EQIP in the United States, the program does not fund infrastructures and targets the development of management methods that, in a manner consistent with the federal approach, could just as well be viewed as educational initiatives and awareness projects.[11] In other words, given the money involved and the educational nature of several of its activities, the environmental dimension of CARD appears rather unintrusive.

It is striking to note the extent to which the federal government avoids command-and-control regulations and relies instead on what I have called the educational approach to addressing environmental problems in agriculture, as well as in other sectors. Part 3 of the new CEPA, which pertains to information gathering, makes it a central task of Environment Canada to "operate and maintain an environmental monitoring system, conduct research and studies and publish information." Knowledge produced in this manner should serve to set "non-regulatory science-based targets or recommended practice."[12] Canadian policy makers appear to assume that, once equipped with scientific knowledge, environmental officials and other environmentally concerned citizens will be efficient at persuading polluters to change their practices.

This focus on education in the agricultural sector amounts to low policy-making performance. As underlined in Chapter 2, convergence on a single approach carries certain risks. The absence of an expert consensus on a single agro-environmental policy approach leaves policy makers with little certainty as to which instrument type produces the best results. Under such circumstances, instruments should be packaged to reflect a mix of approaches in a manner consistent with that observed in France and the United States.

The Ontario Agro-Environmental Effort

A mix of approaches may be more apparent when both levels of government are considered, as in the United States, but a general survey of the provincial situation appears to indicate otherwise. Nolet notes that the provinces rely heavily on voluntary guidelines to address problems of agricultural pollution. He writes that "in most Canadian provinces, the location of farm operations, manure storage, fertilization, etc., are not subject to command-and-control regulations. These aspects are only subject to codes of practice and guidelines."[13] It should be noted that Quebec stands out as an exceptional case since it is the only province that has not hesitated to use intrusive and comprehensive command-and-control regulations and financial incentives in a manner similar to France. I will not say more about Quebec here since I have already dealt with this exceptional case elsewhere.[14] I will instead focus on Ontario, which appears to be more representative of the general provincial trend toward reliance on voluntary, and often informational or educational, measures.

In Ontario the Environmental Protection Act does not apply to animal waste disposed in accordance with "normal" farm practices. Instead, the Ontario Ministry of Agriculture, Food, and Rural Affairs (OMAFRA) and the Ontario Ministry of Environment and Energy (OMEE) produced a voluntary Agricultural Code of Practice to be used as a "guideline" for securing certificates of compliance for buildings and minimum distance separation. More recently, the Agricultural Code of Practice was supplemented with three booklets that provide new guidelines on managing nutrients and minimum distance separations. The Farm Practices Protection Act of 1988 provides producers with protection against nuisance lawsuits over dust, noise, and odours as long as they adhere to "normal" farm practices. Although the Farm Practices Protection Act is subordinated to the Environmental Protection Act and the Ontario Water Resources Act, it does not include additional requirements to be met in terms of environmental protection. Nor does the Ontario act define very clearly what constitutes "normal" farm practices.[15] In fact, this act is more a mechanism for dispute settlement between farmers and their neighbours than an agro-environmental policy instrument.[16]

The Ontario Planning Act, the Ontario Water Resources Act, and the Conservation Authorities Act are also agro-environmental policies but only in a marginal way. The Ontario Planning Act subjects official municipal plans to a provincial policy statement on the protection of prime agricultural land through the inclusion of the Minimum Distance Separation guidelines and the promotion of "normal farm practices." The Ontario Water Resources Act prohibits the discharge of any substance (including agricultural waste) that could impair water quality into any water body or watercourse. Conservation authorities may require permits for work, including agricultural work, in watercourses and flood zones, whereas some municipalities require a certificate of compliance before issuing building permits.[17] In brief, as summarized in Table 6.2, regulatory policy instruments in Ontario display little in the way of intrusiveness or comprehensiveness.

The Environmental Farm Plan (EFP) program is arguably Ontario's most important agro-environmental policy instrument. The Ontario Farm Environmental Coalition initiated the program in 1992, hiring the farmer-controlled Ontario Soil and Crop Improvement Association to organize workshops during which producers are encouraged to conduct broad assessments of their operations. Based on this assessment, an environmental plan is produced and submitted for peer review. Plans can cover a wide range of practices but generally lead to only small improvement projects. Moreover, the financial incentive of a maximum grant of Cdn$1,500 is small compared to the agro-environmental financial incentives granted to farmers in France and the United States. Initially, the EFP was financed by a Cdn$9 million grant from the federal government's Green Plan. In April 1997 it was supplemented by a Cdn$5.6 million grant from the province-based Adaptation Council to continue the program until the year 2000. For its part, OMAFRA provides some agronomic expertise.

Despite the possibility of obtaining a financial incentive, the program appears to be geared toward raising the environmental awareness of farmers, and, in this sense, is not so far removed from the educational approach favoured by the federal government. However, the central role of farm groups in the management of the Environmental Farm Plan program and the minimal role of state agencies appear to be key features in comparison to federal agro-environmental initiatives. Interestingly, farm groups even play a central role in the management of conventional educational programs in Ontario. Ontario's Best Management Practices project, which oversaw the production of booklets, was funded by Agriculture and Agri-Food Canada, managed by the Ontario Federation of Agriculture (OFA), and supported by OMAFRA.

In short, Canada, on the whole, displays low agro-environmental policy-making performance. This situation bears some similarity to France's performance in the 1980s, when that country mostly relied on the CORPEN to educate its farmers. The difference between France and Canada rests in the stagnation of the situation in the latter country. Since the 1980s, Canadian policy makers have trusted farmers as stewards of the land and have taken the position that they need state support only in the form of information and education – a situation that is even more accentuated in Ontario. As a result, between the 1980s and 1990s, Canadian policy makers developed few intrusive command-and-control regulations, and the financial incentives provided are far from what is offered in France and the United States. Such a focus on education surely poses no threat to the economic viability of agriculture, but its environmental effectiveness is questionable. As an illustration, in the summer of 1999, it was reported that only about 5,700 of over 40,000 Ontario farmers had applied for an EFP grant,[18] even though accompanying obligations are minimal. Not even a year later, a major political scandal was caused when the drinking water system of Walkerton, a municipality of rural Ontario, was contaminated with a deadly bacteria, killing seven people. Agricultural run-offs were found to be the immediate cause of the contamination. But in his inquiry report, O'Connor did not hesitate to blame Ontario's environmental policy for the tragedy, stressing that the province's environment ministry needs adequate resources to intervene in rural Ontario.[19] The current discussion surely shows that Canadian agro-environmental efforts are weak in comparison to 1990s France and even to the United States.

Explaining the Canadian Performance

At the heart of this poor Canadian performance lie the actor constellations.

Table 6.2

Canadian agro-environmental policy approaches and the comprehensiveness/intrusiveness of their instruments

Approach	Instrument	Federal	Ontario
Regulatory	Comprehensiveness	Low	Low
	Intrusiveness	Low	Low
Cross-compliance	Comprehensiveness	N/A	Low
	Intrusiveness	N/A	Low
Incentive	Comprehensiveness	High	High
	Intrusiveness	Low	Low
Education and extension	Comprehensiveness	High	High
	Intrusiveness	Low	Low
Self-regulatory	Comprehensiveness	N/A	High
	Intrusiveness	N/A	Low

While the policy network is more problematic in Ontario than at the federal level, the actor constellations, at both levels, are insufficiently diversified to prevent reliance on a single nonintrusive agro-environmental approach.

The Federal Actor Constellation and Network

The Canadian agro-environmental policy network is one of pressure pluralism. Civil society groups, farmers, and environmentalists do not directly participate in policy making. In fact, they appear fragmented and weak in comparison to their American counterparts. The relative decentralization of agricultural and environmental responsibilities to subnational governments has contributed to the fragmentation of groups along provincial lines. On the other hand, the weakness of government agencies ensures a balance of power between civil society groups and the state (see Table 3.1). Neither Agriculture and Agri-Food Canada nor Environment Canada constitute powerful bureaucracies in Ottawa capable of state direction. Within such a policy network, one should expect policy makers to act as brokers between the opinions present within the actor constellation, rather than as leaders.

In comparison to the constellations observed thus far in France and the United States, the Canadian federal constellation appears surprisingly cohesive. In contrast to France and the United States, Canada is a small country with an export economy. The stakes of maintaining the competitive edge of Canadian industries are thus very high, and agriculture is naturally not excluded from this situation. Officials in Ottawa therefore believe that the regulatory burden on business needs to be treated with caution. In 1992, the federal government launched a review of existing regulations, which was particularly aimed at identifying more flexible approaches. A program to reform federal regulation making was then created in 1994. This program requires regulatory agencies to adhere to a strict process in developing new regulations. The process includes: demonstrating that the level of risk is sufficiently high to justify regulation, undertaking public consultations, carrying out a cost-benefit analysis to show that benefits are higher than costs, minimizing the impact on business competitiveness, minimizing the costs for Canadians, and demonstrating that the enforcement capacity is adequate.[20] Lemaire argues that this policy amounts to nothing but a "regulation for regulations."[21] Given the burden that the process represents for regulatory agencies, it is not surprising that federal officials, including those who work for Environment Canada, a weak bureaucracy, will seek other approaches before considering command-and-control regulations. Although some actors in the constellation may still believe tougher regulations would improve the environment, their low feasibility in Ottawa discourages their advocacy.

In addition, other approaches were used in the agro-environmental sector prior to the development of this regulation for regulations. In 1984 the Senate Standing Committee on Agriculture published a report stressing the importance of soil erosion.[22] In response, the federal government, in collaboration with provincial governments, launched a series of programs[23] whose objective was to gather information on practices minimizing soil erosion and then to convince farmers to adopt these practices. Notably, the responsibility for administering these programs fell to Agriculture and Agri-Food Canada, rather than to Environment Canada. Furthermore, these programs were perceived as a success. In fact, a large proportion of Ontario and western Canadian farmers adopted new seeding technologies in order to leave crop residues on their fields to prevent soil erosion. Buoyed by this experience, agricultural officials have every reason to believe that the educational approach is effective in the agro-environmental sector. In other words, the policy experience of the 1980s had a feedback effect well into the 1990s.[24] It is now viewed as the normal policy trajectory to delegate agro-environmental responsibilities to agricultural officials and to rely on the educational approach.

While satisfied with the educational approach, farm groups would still be inclined to accept more generous financial incentives. Yet the largely unfavourable economic context of agriculture pushes agro-environmental aids far behind other forms of subsidies in the policy preferences of Canadian farmers. Unlike in France, for example, agricultural income support programs were cut back to a large extent in Canada in the 1990s, leaving many farmers in a vulnerable economic situation.[25] When government finances improved rapidly in the mid-1990s, the Canadian Federation of Agriculture, the main farm group in Ottawa, began pressing for a reinvestment in income support programs, a reinvestment that appeared even more urgent as the Asian crisis unfolded. But the pressures proved largely unsuccessful since the federal government kept arguing that it did not have the resources to compete with the subsidies offered to farmers in the United States and in the European Union. The government argues that the problem should be tackled at the source and has thus devoted efforts on the international scene to obtaining agreements on the reduction of price-depressing and trade-distorting agricultural aids. As long as the federal government maintains its refusal to refinance income support programs, farmers will feel there is a more pressing problem than agricultural pollution, thus preventing farm groups from making agro-environmental subsidies an important demand. In addition, farmers' dissatisfaction with the federal government renders the imposition of intrusive command-and-control agro-environmental regulations politically difficult, leaving the educational approach as the most feasible alternative.

For all these reasons, the federal agro-environmental constellation is highly cohesive in rejecting command-and-control regulations and embracing education, which is a problematic situation from an agro-environmental performance point of view. Only environmental groups, which are at best on the margins of the constellation, appear to refuse this reliance on a single approach: namely, the educational approach. Problems arise when cohesion prevents problem solving. As argued in Chapter 3, problem solving requires a balance between cohesion and diversity, and the absence of the latter prevents the "power of joint action" from occurring.[26] Moreover, pressure pluralism makes it difficult to achieve governed interdependence, whereby state agencies can play a leadership role based on crucial information provided by civil society groups.[27] Under such network and constellation conditions, it is not surprising that the federal government was unable to achieve a stronger performance.

The Ontario Actor Constellation and Network

The Ontario agro-environmental actor constellation is smaller than its federal counterpart. Not surprisingly then, it is even more cohesive. In Ontario, environmental groups and the environment ministry fall outside the constellation because Ontario's Environmental Protection Act does not include "normal" farming practices within its ambit. Accordingly, core agro-environmental issues have been considered a matter of agricultural policy from early on and thus fell under the responsibility of the Ontario Ministry of Agriculture and Food. Since it was not directly involved, the Ontario Ministry of Environment and Energy did not have to develop an expertise in farming practices. Because the Environmental Protection Act does not provide the OMEE with agro-environmental responsibilities, environmental groups in Ontario have rarely been asked to contribute to agro-environmental policy making and, just like the ministry, possess an underdeveloped expertise in agriculture.[28] Therefore, the Ontario constellation excludes those actors most likely to demand stringent regulations.

Since "normal farm practices" and their environmental effects are considered a matter of agricultural policy in Ontario, the province's predisposition has been to mandate the agriculture ministry to deal with any agro-environmental issues. In 1972 an agreement with the United States committed Canada to improving the quality of water in the Great Lakes, and agriculture was identified early on as one of the sources polluting the lakes. Adding to this concern, the Senate of Canada, as underlined above, began pointing to problems of soil erosion in the 1980s.[29] As a result, agreements between the Ontario and federal agriculture ministries led to the launch of several federal-provincial soil conservation programs in the province, which created a self-regulatory capacity among Ontario farmers. The Soil and Water Environmental Enhancement Program (SWEEP), initiated in 1986 and terminated in 1992, drew heavily on farmer participation.

Even more importantly, Land Stewardship I and II (1987-94) involved the hiring of a farm organization, the Ontario Soil and Crop Improvement Association (OSCIA), to administer the programs. With this self-regulatory capacity thus developed in the OSCIA, it is not surprising that Ontario's farmers hired this same organization to deliver the Environmental Farm Plan program.

In short, the Ontario constellation comprises a coalition of the major farm groups (discussed in greater detail below) that possess a self-regulatory capacity and the Ontario Ministry of Agriculture, whose traditional role is to administer farm programs. Such a constellation will be highly cohesive since it is primarily concerned with the promotion of farmers' interests. Attributing an agricultural agenda to the Ontario agro-environmental constellation, one interviewee presented the situation in the following terms:

> The environment agenda is often ecosystem-based or watershed-based. What you do on farms has to be looked at in terms of the receiving body of water, looking to the common good of the watershed as a whole. By doing this, there may need to be some constraints on farm productivity ... that could be offset by a grant, but we might expect the farmer to sacrifice productivity. In contrast, the agriculturalist agenda focuses more narrowly on single farm operations, trying to address enhancement of efficiency on the farm. Things that are done for the environment should be done at minimal cost to the farmer.[30]

It may nevertheless be surprising that the Ontario Ministry of Agriculture adheres so completely to such a view as to endorse self-regulation. State agencies generally claim that they are the legitimate protectors of the public good, and therefore they do not easily delegate their responsibilities to less legitimate civil society actors. I argue that self-regulation is possible in Ontario because the structure of the relationship between farm groups and OMAFRA – in other words, the policy network – has become one of clientelism (see Table 3.1).

When agro-environmental issues began to appear on the policy agenda in Ontario in the 1980s, they entered a pressure pluralist policy network. On horizontal issues, Ontario farmers have been represented by three general farm organizations – the Ontario Federation of Agriculture, the Christian Farmers Federation of Ontario (CFFO), and the National Farmers Union (NFU) – as well as by a large number of commodity groups. Skogstad notes that traditionally their relationships with the agriculture ministry were informal, with their positions reflecting differences on what constitutes acceptable state intervention.[31] In the early 1990s she concluded that the policy network featured a mix of the usual pluralist advocacy politics and some joint action.[32]

The election of a New Democratic Party (NDP) government in September 1990 pushed Ontario's farm organizations in a direction that ultimately led to clientelism. When the NDP proposed to enact an Environmental Bill of Rights that would provide every citizen with the right to a clean environment, it became an important concern for the farming community. As one agricultural leader reported: "Farmers own an important share of rural Ontario; so if the province was to have an Environmental Bill of Rights, the community impacted the most would probably be farmers."[33]

To address this concern, at the beginning of the NDP mandate the newly appointed minister of agriculture formed an advisory committee of senior agricultural leaders that was to meet roughly once a month. This committee was instrumental since it provided a forum favourable to the development of a common view among agricultural leaders on agro-environmental policy. In the beginning, the work of the committee consisted of learning from a number of government agencies about their agro-environmental projects and then advising the minister on their appropriateness. In the process, however, these interchanges contributed to farm leaders reaching the conclusion that they were "spending too much time reacting to the environmental agenda of other groups, but not enough time addressing [their] own."[34] Accordingly, in July 1991 some forty agricultural leaders met in Guelph to consult on a possible "environmental agenda" for the farming community. During the meeting, the leaders decided to form a working group mandated to define such an agenda. Composed of officials from the OFA, the CFFO, and several commodity organizations, this group prepared a document entitled "Our Farm Environmental Agenda," which was published in January 1992.[35] It was signed by a coalition of four groups that shortly thereafter formed the Ontario Farm Environmental Coalition (OFEC), including the OFA, the CFFO, and two coalitions of commodity groups: AgCare (producers of field crops) and the Ontario Farm Animal Council (livestock producers).

Normally, similar coalitions, such as the ones formed in the United States as a means for civil society actors to increase their influence, are not expected to have the same capacity for sustained policy participation as a peak association like France's FNSEA. Such coalitions draw their resources from separate contributions made by members whose policy preferences may be quite changeable, as we saw in the previous chapter.[36] However, in the Ontario case an important change was introduced into the agricultural associational system shortly after the OFEC was formed. In 1993 the Ontario government passed the Farm Registration and Farm Organizations Funding Act, which required farmers to pay dues to an accredited general farm organization: either the OFA or the CFFO.[37] This law had two effects. First, it gave these organizations a kind of official public status and legitimacy that they did not previously have. Second, it added to their resources, especially the resources of the OFA, the larger of the two accredited groups. With this added capacity, the coalition gained enough permanence and strength to subordinate OMAFRA in a clientelist network within the agro-environmental actor constellation. Under such circumstances, it was easy for the OFEC to take a leadership role on other agro-environmental policies such as water quality and nutrient management.[38] Given this situation, OMAFRA possesses little capacity to circumvent farmers' plans in the agro-environmental sector.[39] The ministry is reduced to providing services of minor strategic importance within the framework set by the Environmental Farm Plan.

In fact, no state actor can prevent farmers from realizing their objectives in such a clientelist network. Elected in 1990, the NDP government began pushing to open the agro-environmental actor constellation to broader environmental interests. Already in the 1980s, the environment ministry had, in a series of studies, identified agriculture as a major cause of water contamination. Accordingly, an important agricultural component was included in the Clean Up Rural Beaches (CURB) program, a ten-year, Cdn$60 million environmental program approved by the NDP government in 1991. It was expected to address the problem of beach closures due to bacterial contamination in rural areas. Through CURB, the Ontario Ministry of the Environment began to emphasize the need for changes in farm practices. As a result, the NDP began to develop a certain organizational capacity to address agro-environmental problems within the government's main environmental agency.

However, the OMEE was never able to penetrate the agro-environmental actor constellation. Arguing that agro-environmental problems were well addressed by the OFEC and OMAFRA, the newly elected Progressive Conservative government stopped the process in its tracks. Shortly after the 1995 election, the government announced the premature termination of CURB. It then rapidly proceeded to cut 33 percent of the OMEE staff, which, according to interviewees, effectively deprived the ministry of any limited capacity it had to deal with agricultural pollution, outside of pesticides. While the election of the Conservative government consolidated the clientelist network, the rapid reversal of the NDP's attempt to widen the agro-environmental actor constellation was no doubt made possible by the presence of the clientelist network. Clientelism deprives the state of its leadership role, and governed interdependence thus becomes problematic and the transformative capacity of policy initiatives is reduced.[40] With such a policy network in place, self-regulation in the agro-environmental sector in Ontario is relatively safe.

Naturally, the Walkerton tragedy will not be without consequences. Indeed, in response to O'Connor's inquiry report, the Ontario government adopted in June 2002 the Nutrient Management Act. Enabling the development of the province's regulatory standards for nutrient management generally, the act's central objective is to gradually make it mandatory for farms to have nutrient management plans. In the process of standard development, the act promises to involve the Ministry of the Environment, thereby diversifying the actor constellation. However, one may see the introduction of the act into the legislative assembly by the minister of agriculture as a sign of continued predominance by the Ministry of Agriculture. And if the analysis presented in this book is correct, it is not even clear that, in the short term, the Ministry of the Environment would possess the necessary agro-environmental expertise to make a significant contribution within the constellation. Likewise, the subordination of the agricultural ministry to farm interest groups in a clientelist network will not be changed overnight. Possibly not by coincidence, the agriculture minister insists on existing self-regulatory norms as a source of inspiration for the regulatory framework the act is enabling: "Many guidelines and other reference documents have already been developed which could provide a good basis for these standards. Examples include the Ontario Farm Environmental Coalition's Nutrient Management Strategy, Environmental Farm Plan and many Best Management Practices."[41] It is early to make a definitive judgment on the Nutrient Management Act. However, there are indications that even in the face of a tragedy as significant as Walkerton, the clientelist network prevailing over Ontario's agro-environmental sector will be altered only in the long term.

Conclusion

Canadian agro-environmental performance contrasts sharply with that observed in France and the United States. The Canadian actor constellations at the federal level and in most provinces decided in the 1980s to rely mainly on education and continued to do so into the 1990s. While education might contribute to reducing agricultural pollution, the opinions of epistemic communities are too widely divergent to be certain whether these educational policy instruments will suffice. Under these circumstances, relying on a mix of instruments is preferable, as I have argued in Chapter 2. Education is by nature a nonintrusive approach. While it may in some circumstances serve to convince farmers to change their farm practices, it creates no obligation. Farmers who are unconvinced, or who accord a greater weight to their interests than to their beliefs, can resist a change in their practices without suffering any consequences.

The cause for this stagnation in Canada can be found in the evolution, or lack of evolution, of the actor constellations and policy networks. First, the actor constellations are surprisingly cohesive. At the federal level, this cohesion can be traced back to a now-institutionalized bias against command-and-control regulations and to feedback from educational soil erosion programs developed during the 1980s. In Ontario, the cohesion is mostly related to the size of the actor constellation. In fact, the constellation excludes the Ministry of the Environment and environmental groups, a situation that has a history dating back to the failure of Ontario's Environmental Protection Act to include normal farming practices within its ambit. Second, the agro-environmental policy network in Ontario has evolved from pressure pluralism to clientelism. Within such a network environment, state agencies are deprived of their capacity to change a policy approach that their "clients" consider adequate.

Interestingly, these constellation and network situations diverge somewhat from the theories discussed in Chapter 3. First, the evolution away from pluralism and toward clientelism in Ontario has little to do with the new politics of the welfare state. While the development of an agricultural policy in Ontario has certainly encouraged the establishment of farm groups, clientelism has arisen from a succession of single events: the election of the NDP; the creation of a committee of agricultural leaders; the formation of the OFEC; the adoption of a law on the financing of farm organizations; and the election of the Conservative Party, which lead to consolidation of the clientelist structure.

Internationalization is a negligible factor in the explanation of the cohesion of the Ontario constellation.[42] Internationalization, however, which I have argued in Chapter 3 takes on an intergovernmental character in the North American context, may have played a larger role at the federal level. When developing command-and-control regulations, policy makers appear to be aware that businesses are sensitive to added costs since they belong to a small, open, export economy. From this angle then, internationalization presents itself more as a belief than as a structural development shaping policy networks in a significant manner. First, internationalization has failed to reorient the federal agro-environmental policy network away from pressure pluralism and toward state direction (see Table 3.2). Second, environmental groups and the federal environment ministry were not weakened by internationalization since they were already weak in the first place. The federal environment ministry, Doern and Conway argue, has never been able to make a convincing case in Ottawa that the environment is more than an "amenity concern."[43] Third, the fact that a large portion of Canadian agricultural production is exported does not constitute a new development. Concerns of the export sector have always occupied an important place in actor constellations responsible for agriculture related issues. In short, the international competitiveness of the farm sector may occupy an enviable position in the discourse of actors in constellations, but internationalization is far from having the structural impact some theories suggest, even in a small, open economy like Canada's. I will take a closer look at these theoretical discussions in the next and final chapter.

7
Misplaced Distrust

This book is about output-oriented legitimacy, or the extent to which policy solutions should inspire citizens' trust in governance arrangements. This question has been addressed by breaking it down into three steps. The first step involved identifying a politically challenging problem with performance-generating solutions. The selected problem was agricultural pollution, and the development of intrusive, comprehensive, and economically sensitive solutions to agricultural pollution was used to indicate a high level of performance. Where such high performance levels were found, it was suggested, policy makers are likely to be deserving of trust.

The second step consisted of verifying the extent to which strong policy-making performance was theoretically attainable. It was argued that cohesive, yet sufficiently diversified, actor constellations whose interactions were structured by corporatist policy networks would provide conditions conducive to strong policy-making performance. However, theoretical developments in the field of comparative public policy suggest that actor constellations and policy networks might evolve away from such conditions. In other words, academically rooted theories may feed into citizens' mistrust of governance structures.

The third step was to verify whether this mistrust with regard to policy networks' capacity to design adequate environmental policies was empirically valid. Surprisingly, a careful examination of environmental policy development for the agricultural sector in France and in the United States revealed levels of policy-making performance that should inspire trust. Inadequate policies were observed only in Canada. Moreover, it was shown that Canada's performance has little to do with factors that, in theory, are usually associated with poor performance.

This chapter will revisit the actor constellations responsible for the performance levels of France, the United States, and Canada to demonstrate that the factors often associated with poor performances may, under specific conditions, be conducive to strong performance. Obviously, trust-inspiring improvements in Canada, the United States, and France do not demand the changes, such as bureaucratic cutbacks, typically advocated by several contemporary thinkers.

This chapter is divided into three sections. First, the performances of France, the United States, and Canada will be compared. Second, a systematic analysis of these three countries' actor constellations and policy networks will be presented. Finally, the confidence-eroding theories will be revisited to show that distrust is often misplaced.

Three Levels of Performance

An adequate policy for addressing agricultural pollution must: (1) recognize the problem; (2) be designed to include measures intrusive enough to change the behaviour of farmers; and (3) cover a comprehensive enough range of practices so as to prevent pollution transfers across media. An adequate policy should also: (4) attempt to avoid endangering the economic viability of agriculture insofar as possible. Of the three countries examined here, France comes closest to meeting all of these criteria.

At the outset, it should be emphasized that even though all three countries have not adopted equivalent measures for addressing agricultural pollution, they have all moved the problem onto their political agendas. Agricultural pollution, understood as a conservation problem, has been on the agenda of the American federal government since the 1930s. The meaning of conservation has broadened since then to gradually include problems associated with livestock production. Toward the end of the 1980s, agricultural pollution even reached the agenda of state-level governments. In France agricultural pollution has been on the political agenda ever since the early 1980s. As a sector viewed as being in need of modernization, agriculture in France has been accorded special treatment since the 1960s. As a result, it took a special event – the publication of the Hénin Report – to place agricultural pollution on the political agenda of that country. Interestingly, agricultural pollution also made it onto the agenda of the European Community after the adoption of the Single European Act in 1986. In Canada, despite the few policy efforts of the federal and provincial governments, agricultural pollution has nevertheless occupied an important place on the governments' agenda. Defined in terms of soil erosion following the publication of *Soil at Risk*, a Senate report,[1] the federal and provincial agendas evolved to comprise water pollution in the 1990s.

Differences between these three countries become more apparent when the intrusiveness of the policy instruments adopted to tackle agricultural pollution are examined. France and the United States were able to adopt stringent regulations and provide financial incentives generous enough to expect farmers to modify their practices. On the other hand, Canada chose not to adopt command-and-control regulations or to provide the financial aid likely to incite similar changes in farm practices. Canada has instead chosen to rely on voluntary programs, most of which are of an educational nature.

It is interesting to note that France did not immediately opt for intrusive instruments when agricultural pollution hit the political agenda. In the 1980s, French policy makers decided to form the CORPEN, a committee responsible for educating farmers, and explicitly rejected command-and-control regulations. It was only early in the 1990s that intrusive instruments such as the Nitrate Directive, the extension of the water policy to agriculture, and the PMPOA appeared. Likewise, the American federal government, accustomed to a conservation approach, resisted intrusiveness. One might argue that in terms of its generosity, the CRP is an intrusive measure, but interviewees made it clear that the program in the 1980s was not an entirely environmental program but rather was geared toward income support during a period of price depression caused by large surpluses. However, governments in US states where agricultural pollution was a major problem did adopt intrusive policies in the form of command-and-control regulations.

Proposing a general assessment of the comprehensiveness of the policies adopted by each country is a more difficult exercise, as feasibility varies according to instrument type. For example, a country like Canada, which essentially relies on educational policy instruments, might be expected to do quite well since educational instruments, while naturally nonintrusive, are often comprehensive. On the other hand, one might expect a country like France, where regulations are important, to be doing poorly since intrusive regulations are made comprehensive at a very high cost. In any case, even with these differences in feasibility, France still managed to perform better than Canada and slightly better than the United States. Even though command-and-control regulations in France have become only moderately comprehensive, the country has developed a wide range of instruments to cover an impressive number of farming practices.

As mentioned above, the American CRP is viewed as a limited environmental policy instrument – at least in its original form – because it targets income support in addition to environmental protection. A program with two such conflicting goals is less likely to achieve either of them in a satisfactory manner than is a program with only one goal. Nevertheless, the development of EQIP and of the Conservation Reserve Enhancement Program, which allocates some funds to livestock farmers for the adoption of practices with lower environmental impacts, renders conservation measures more comprehensive for what seem to be environmentally motivated reasons. In short, while not covering a range of practices as broad as those in France, the United States's performance is not too bad in terms of comprehensiveness.

Incentive programs in Canada are also rather comprehensive, but they compare poorly with those of the United States since improvements in comprehensiveness have not engendered higher investments. Canadian programs allocate so little money that their comprehensiveness is likely to disperse resources rather than to direct them toward the most serious problems. Furthermore, the very few regulations that do exist in Canada are not comprehensive; only educational measures appear to be comprehensive, but, as explained above, it is difficult to accord much merit to this.

Environmental measures in Canada impose so few constraints on farmers that it makes little sense to question the extent to which policy makers were capable of taking into account the economic stakes related to agriculture when developing an environmental policy for the sector. The situation is different in France and the United States, and herein lie some of the important differences between the two countries. In France, farmers are appropriately compensated through the PMPOA for modifying their practices. While it is difficult to assess whether the PMPOA constitutes too little or too much compensation, it definitely represents an effort to protect the economic viability of the French agricultural sector in the face of serious environmental obligations for farmers. In the United States, generous financial aid programs also exist to compensate farmers who adopt practices with lower environmental impacts. However, the most intrusive command-and-control regulations requiring such changes are state regulations, whereas the main compensatory programs are federal conservation programs. While these programs have come to cover a wider range of practices, they are not coordinated with the various state policies. As the comprehensiveness of US federal conservation measures increases, state requirements and federal compensation may coincide more frequently. However, given the variety of state command-and-control regulations, it is unlikely that compensation will ever become as consistent with the actual changes made on farms as it is in France. In other words, the American farmers who need help most to comply with regulations might not actually be the ones receiving the aid.

The absence of coordination between state regulations and federal incentive programs, which disconnects the needs from the financial resources, is not without economic consequence. The danger here, in terms of agriculture's economic viability, depends on the extent of the consistency between regulatory requirements and compensation. Again, as federal aid programs become more comprehensive, federal compensation programs and state regulations may draw closer together, but the absence of coordination will remain of dire economic consequence for some unfortunate farmers. Therefore, even if the absence of economic coordination were to have only a small economic impact for the country, for individual farmers the American situation is not as ideal as that seen in France, where compliance with regulations entitles farmers to compensation.

In spite of having placed agricultural pollution on its political agenda, Canada has demonstrated a poorer performance than either France or the United States on all counts. Canada has failed to adopt intrusive and comprehensive environmental instruments for agriculture and thus has not had to develop a means to address the economic implications of its policy. In contrast, France and the United States have adopted an interesting mix of intrusive command-and-control, as well as financial aid, instruments. While both of these countries have adopted policies indicative of strong environmental policy capacity in agriculture, France seems to have done a little better than the United States. France covers a slightly more comprehensive range of farming practices because of the variety of policy instruments it has adopted. More significantly, however, France has been able to provide better economic protection to its farmers since financial aid is coordinated with regulatory requirements.

The empirical chapters of this book have argued that performance changes or the absence of changes in performance in each of these three countries are connected to their respective actor constellations and policy networks. In the comparative analysis of the three actor constellations and networks that follows, it should become quite clear that different actor constellations and networks produce different performance levels.

Three Actor Constellations and Three Policy Networks

Following Scharpf, Chapter 3 presented the argument that actor constellations are most likely to produce problem-solving solutions when they are cohesive enough to allow dialogues or negotiations to take place. However, they also need to include sufficiently diversified and complementary expertise if the "power of joint action" is to occur.[2] Weiss adds to this perspective, saying that the structure of the relationships between the actors matters. In line with Scharpf, Weiss argues that a relationship structure that institutionalizes the interactions between civil society and state actors, who possess interdependent resources, is likely to produce better policies. However, she adds that this interdependence must be governed by a relatively autonomous state. The policy transformations necessary in a changing environment are most likely when state actors are capable of elevating themselves above the narrower interests of civil society actors.[3] I have argued that these conditions are captured by what Coleman calls a "corporatist policy network."[4]

These types of performance-generating conditions are currently most closely encountered in France. Thanks to the Nitrate Directive, the French agro-environmental actor constellation now includes the Ministry of the

Environment, in addition to farmers' representatives and the French agriculture ministry. Composed exclusively of agricultural actors in the 1980s, this actor constellation was not able to achieve the power of joint action. In addition to the fact that the actors in this constellation had little interest in developing an environmental policy for agriculture, they were not accustomed to thinking in terms of the environmental impact of agricultural practices and possessed no experience in environmental policy development. The inclusion of the environment ministry within the constellation provided the needed complementary expertise. Together, the agricultural actors and the environmental actor developed policies that none of them could have developed on their own.

It is not insignificant that the environmental actor included was a ministry rather than an interest group. Organized environmental groups in France are often too weak and possess too little capacity to effectively participate in policy development. Such weakness confines them to a policy advocacy role. Under such circumstances, environmental ministries can act as efficient substitutes to ensure the inclusion of environmental concerns in policy decisions.[5] Environmental groups have participated constructively in environmental policy development elsewhere in the world,[6] but the substitution of their leadership by that of a ministry provides for safe dialogue conditions. Since they belong to a state apparatus, ministry officials are accustomed to interacting with other ministries on a variety of subjects. These interactions work to refine their openness to different points of view and their capacity to compromise. While the danger of cooption to broader state objectives overlooking environmental protection always exists, the proximity of environmental officials to an environmentally concerned public should act as a safeguard against such a danger. In short, the entry of the French environment ministry into the actor constellation provided complementary expertise without threatening the cohesion necessary for establishing dialogue or negotiation among the actors. These are conditions conducive to achieving the power of joint action.

French policy development in agriculture has traditionally been undertaken through a corporatist network that places the FNSEA, the main farm organization in France, into a veto player position. The entry of the environment ministry into the constellation did not act to erode this relationship structure between the state and civil society and thereby prevented a policy shift in the environmental direction harmful to farmers' economic interests. While the environment ministry was able to exercise some leadership, new environmental policy measures had to be discussed among state officials and farmers' representatives. As explained in Chapter 4, these discussions did not take the form of a problem-solving or truth-seeking process in the 1990s since distributive issues could not easily be set aside. Nevertheless, the negotiations arising out of the corporatist network displayed welfare-enhancing qualities. The compensation offered to farmers under the PMPOA proved to be enough of a side payment to integrate agriculture into the water regulatory framework. In addition, after a decade of close collaboration between environmental and agricultural actors within a corporatist network, the conditions appear increasingly favourable to problem solving. I have in fact argued that, more than the other programs of the 1990s, the Contrat territorial d'exploitation (CTE) is not easily understandable through the lens of a distributive contract. Problem solving, which involves truth-oriented dialogue, seems to be a better interpretation.

To fully grasp the American situation in the agro-environmental sector, both the national and the subnational levels must be accounted for. In other words, in contrast to France, actor constellations at both the federal and state levels shape agro-environmental policy development. Taken alone, each of the actor constellations would appear inadequate to produce high-performance environmental policies for agriculture. On the one hand, the state-directed federal policy network places the USDA and its conservation approach in such a position as to prevent the pooling of complementary expertise. On the other hand, state-level pluralist networks, which disable governed interdependence, allow for a radical response to public mood swings and other environmental pressures. In other words, the federal policy network is geared toward reliance on a single (and perhaps not the most promising) approach, whereas the state networks might enable the adoption of environmental command-and-control regulations in such a manner as to endanger the economic viability of farming. However, when these constellations are taken together, some merit can be seen in the American situation. Again, the financial aid provided by federal incentive programs may serve to compensate for the adoption of state-level command-and-control regulations.

Nevertheless, this situation is not equivalent to that found in France. While it is impossible to assess precisely the effect of the PMPOA in France, the manner in which it was decided suggests welfare gains. First, the participation of the FNSEA in the negotiations guaranteed that payments were treated explicitly as compensations to farmers who comply with the stringent regulations. Second, the participation of relatively autonomous state officials in the negotiations should have served to ensure that the cost of integrating farmers into the water regulatory framework was no higher than the gains. In contrast, the disconnection of American decisions with respect to regulatory requirements and financial compensation, which can be attributed to the presence of two distinctive constellations, does not offer similar guarantees. Given this type of disconnection in the United States, it is very likely that compensations are either too high or too low to prevent economic losses or that they simply go to the wrong farmers. In short, while the environmental approaches adopted by federal and state constellations appear to present interesting complements, the absence of policy coordination renders the American situation less desirable than that found in France.

Stalled at a poor performance level, Canada possesses the least desirable actor constellations and networks. In contrast to the constellations observed in France and the United States, the actor constellation at the federal level in Canada has crossed the line beyond which cohesion becomes a problem. In contrast to the United States, no environmental group active on agriculture has been able to establish itself effectively on a national scale. Environmental groups in Canada are severely fractured along provincial lines. While the group situation is not so dissimilar to that of France, the Canadian state, unlike the French state, does not possess an environment ministry capable of compensating for the absence of environmental groups in the federal constellation. Because it is relatively weak, the Canadian federal environment ministry is incapable of compelling support for the environmental solutions that epistemic communities around the world would consider adequate for agriculture. Interestingly, Canadian farm groups and the agriculture ministry are only slightly more resourceful than the environmental actors; however, the agricultural actors benefit from a discourse on the international competitiveness of business that is popular among federal policy makers. In the name of business competitiveness, federal policy makers have even institutionalized a disincentive for the development of new command-and-control regulations, environmental and otherwise. In short, the Canadian federal actor constellation displays a high degree of cohesion around ideas that coincide with agricultural interests. Environmental officials who have not been coopted by federal policy makers, and whose loyalty is primarily with environmental protection rather than with global competitiveness, find themselves, at best, at the margins of the actor constellation. Such cohesion, I have argued, prevents the power of joint action. Deprived of sound knowledge and expertise in environmental policy making, the constellation has favoured an overwhelming reliance on a single nonintrusive policy approach: namely, agro-environmental education.

In contrast to those in the United States, the actions of subnational governments in Canada, with the exception of Quebec,[7] have offered a poor complement to the federal policy line. A close examination of Ontario has revealed not only that the province's constellation offers no promise in terms of establishing a balance between cohesion and diversity, but also that the provincial clientelist policy network places the state in a subservient position to farmers. Preferring to exercise full control over environmental policy making for agriculture, farm groups, through the OFEC, call on Ontario's agriculture ministry to provide services only of minor strategic importance. Needless to say, such a network curtails all efforts to create a situation of governed interdependence. The NDP government attempted to increase the leadership of the province's environment ministry in agriculture, but these efforts were quickly defeated. In his inquiry report on Walkerton, O'Connor also demands greater leadership from the environment ministry in rural Ontario.[8] Whether the agro-environmental policy network is successfully changed to accommodate this demand remains to be seen. For the moment, in sharp contrast to the situation in France, no state actor has the capacity to elevate itself above the particular interest of Ontario's farming community. Clientelist networks probably do not dominate provincial politics all over Canada, but evidence indicates that provincial constellations and networks outside of Quebec are ill-equipped to produce adequate environmental policies for agriculture.

In summary, high policy-making performance depends heavily on two factors: (1) an actor constellation that offers a balance between cohesion and diversity; and (2) a policy network in which state actors remain capable of exercising leadership while also remaining close to civil society groups efficient enough to provide crucial information for policy development. Such an actor constellation and a policy network have been conducive to high policy-making performance in France. Similar policy-making performance has not been achieved in the United States because the federal state-directed policy network has served to disconnect government officials from civil society. This situation has been partially corrected by state-level pressure pluralist policy networks that, while disabling leadership by government actors, render policy makers highly responsive to swings in public opinion. Therefore, the central problem in the United States is coordination between the activities of state-level and federal actor constellations. Performance failure is sharpest in Canada, where not only are policy networks inadequate, but the actors admitted into the constellations have to hold highly cohesive ideas. Such constellations are deprived of the complementary expertise necessary for achieving the power of joint action or, in economic terms, for producing welfare-enhancing policies. It is interesting to note that, unlike cohesion, diversity within the constellations of the three countries has never reached a point beyond which dialogue becomes impossible and performance is thereby compromised (see Table 7.1).

Theoretical Sources of Distrust Revisited

Critics of both the Left and the Right have said that policy makers respond to an incentive structure that inhibits the development of policies capable of resolving collective problems. Critics of the Left have pictured policy makers as being too deferential to economic interests, whereas critics of the Right present policy makers as being too preoccupied with their own concerns to effectively respond to society's concerns. Such critics have fostered a profound distrust among citizens toward policy makers. To assess the extent to which such mistrust may be justified, I have examined policy responses to the politically challenging problem of agricultural pollution, and I have found, more often than not, relatively promising policy responses. This is doubly surprising, since the literature review provided in Chapter 3 illustrates that this crisis of confidence is, in theory, on solid ground. Naturally, in light of the empirical evidence, these pessimistic theories need to be revisited.

Agenda Setting

A first theoretical source of distrust was found in the literature on agenda setting. More conventional perspectives of agenda setting suggest that problems are considered by policy makers only when they are viewed as serious enough. However, it has been argued, following Kingdon, that agenda setting is not such a linear process since it also involves random elements.[9] In fact, Kingdon suggests that agenda setting depends on the ability of policy entrepreneurs to recognize and act upon the opening of policy windows. Policy windows open when the problem, the solution, and the political streams come together, or, in other words, when the political conditions become favourable to addressing a given problem for which a compelling solution exists. The three streams, Kingdon adds, are largely independent, giving rise to the idea that agenda-setting processes are characterized by randomness.

Howlett contends that the unpredictable character of Kingdon's process has been exaggerated.[10] Nevertheless, as I have argued in Chapter 3, Howlett's "discretionary window," "spillover window," and "routine window" offer a no more desirable image of the policy process in terms of policy-making performance. Citizens who demand proper governance do not want agenda setting to depend on the discretion of one person, or on spillover from other sectors, or on political or administrative routines.

Table 7.1

Variations in constellations and networks

Country	Constellations	Networks
France	Diversified	Corporatist
United States		
Federal	Diversified	State-directed
States	Diversified	Pressure pluralist
Canada		
Federal	Cohesive	Pressure pluralist
Provinces	Cohesive	Clientelist

They want policy networks to react rapidly and even to anticipate problems – that is, to consider actions before the problems turn into disasters. Howlett's window openings are disconnected from problem severity and would render this type of anticipation unlikely.

Assessing the efficacy of agenda-setting processes presents certain difficulties. First, it requires that problems deserving the attention of policy makers be distinguished from those that can wait. Second, it demands a decision as to when problem severity marks the passing from anticipation to reaction. Naturally, these distinctions always involve a certain element of arbitrary judgment. This being said, I have argued that the evidence is sufficiently worrisome to suggest that agricultural pollution deserves the attention of policy networks, and the empirical chapters have shown that the problem hit the government agenda rather early on. Therefore, the process of agenda setting observed in France, the United States, and Canada does come close to being the straightforward, linear, and reassuring image projected by traditional agenda-setting theories. The random and even more predictable openings of policy windows, as well as the ability of policy entrepreneurs, have not exercised a significant influence in positioning agricultural pollution on the political agendas of the three countries.

In fact, it seems remarkable that agricultural pollution arrived almost simultaneously on the political agendas of France, the United States, and Canada. In France, as early as the 1980s, agricultural pollution was looked at in terms of the impact of modern farming practices on soil and water. During the same period, the problem also hit the political agendas of the United States and Canada but was more narrowly viewed as soil erosion. By the end of the 1980s, however, a broader understanding of agricultural pollution was on the political agendas of the governments of all three countries. These early differences between how the problem was viewed by France, on the one hand, and by Canada and the United States, on the other, coincide with differences in terms of problem severity. France has experienced more severe agricultural pollution problems with respect to soil and water than have the other two countries. Consequently, adopting a broader view of agricultural pollution was more compelling for French policy makers than it was for their North American counterparts. This is not to suggest that random elements never play a role – that is, that the almost simultaneous setting of the three countries' agendas completely discards randomness. This evidence nevertheless raises questions about the complete independence of the political, policy, and problem streams.

Some would insist, however, that the almost simultaneous setting of the three countries' agendas is better explained by transnational problem construction, notably as achieved by epistemic communities, than by comparable severity in problems.[11] To such a view, I would add a serious caveat based on the evidence presented in this book: Epistemic communities do not entirely construct problems out of nothing; agendas are not so far removed from concrete reality. Agricultural pollution, whether it is soil erosion or bacteria in water, is concrete and clearly consequential for resource depletion, health, and the economy, and thus it cannot be manipulated rhetorically as perhaps can other problems. Therefore, an adequate agenda-setting theory for a down-to-earth issue such as agricultural pollution should accord priority to the seriousness of the problem. And such a theory would naturally promote a brighter outlook on policy-making capabilities.

The New Politics of the Welfare State

The emergence of the new politics of the welfare state, whereby programs that concentrate benefits act to form networks of interest groups and of state bureaucracies, constitutes a second theoretical source of distrust. Pierson contends that welfare state programs, because they tend to concentrate benefits, have given rise to powerfully organized networks of support capable of locking in policy trajectories.[12] Such policy-rooted networks indeed tend to produce increasing returns since they involve high fixed costs, learning effects, coordination effects, and self-fulfilling expectations.[13] In addition to creating conditions that give rise to this particular form of collective action, existing welfare state policies provide low-cost "mental maps," or logics of appropriateness, to filter new information about complex problems, about policy solutions for which an implementation expertise exists, and about a set of expectations to which actors can adapt. When combined with the short time horizon of politics closely associated with elections, the increasing returns associated with existing welfare state policies provide solid disincentives for breaking the policy trajectory.

As explained in Chapter 3, agricultural policy development is closely associated with the construction of the welfare state. Thus one might expect the path-dependency logic of Pierson also to apply to this sector.[14] As with most social policies, existing farm programs tend to concentrate benefits in a well-defined group of citizens – namely, farmers – who have relatively powerful organizations for collective action. Moreover, farm groups have often established a close relationship with agricultural state bureaucracies experienced in providing production-enhancing services to farmers. In fact, production enhancement constitutes the central expectation associated with welfare state development in agriculture. Given the general esteem in which citizens hold farmers and the low-cost food that welfare state agricultural policy affords, the electoral cost of a break in the agricultural policy trajectory associated with welfare state development could be high.

It was within this type of agricultural policy context that problems of agricultural pollution appeared on the agendas of the French, American, and Canadian governments in the 1980s. Following Pierson, it was reasonable to expect policy makers to examine the problem as a threat to production – a perspective close to that of the epistemic community of agricultural system analysts. Not only are agricultural bureaucracies equipped to address agricultural pollution in this manner, but such an approach to the problem can easily obtain the support of farm organizations. Conversely, the benefits associated with the stricter reduction of most agricultural pollutions are dispersed rather than concentrated, and thus collective action on the environmental front is curtailed more often than not.

With the exception of Canada, however, agro-environmental policy development has broken with the welfare state agricultural policy path. In France and the United States, the ideas that have inspired environmental policy development often conflict with production-enhancing objectives for agriculture. Quite clearly, actions such as integrating agriculture into the French water management regulatory framework or proclaiming a moratorium on the growth of the North Carolina hog industry target pollution reduction, not production enhancement. The French Contrat territorial d'exploitation even squarely breaks with the productivist paradigm. The findings presented in this book suggest that the structural conditions created by past policy decisions are complex.

Pierson's theoretical proposals might underestimate the extent to which policy development tends to cut across several sectors. Developing a policy to tackle agricultural pollution, as I have shown, impinges upon agricultural policy and also upon environmental policy, which is a sector largely disconnected from welfare state construction. As a result, developing a policy for agricultural pollution requires managing potential conflicts between what appears to be appropriate in terms of agricultural policy and what appears to be appropriate in terms of environmental policy. Assessing breaks or continuation with policy trajectories may then depend largely on the alignment between the sector to which one belongs and the logic of appropriateness that wins out. While the development of environmental command-and-control regulations can appear to farmers as a break with past policies, it may appear as a continuation for environment officials who have developed similar regulations previously in other sectors.

In any case, when policy development cuts across several sectors, actor constellations tend to be more diversified than is assumed by Pierson. In two of the three countries examined here, environmental policy development for agriculture involves constellations composed of environmental and agricultural actors whose cognitive orientations may be closely associated with past policy decisions, but these past policy decisions are markedly different. Environmental actors support command-and-control regulations that, as an instrument type, have a long history in the sector, while agricultural actors support financial incentives that have an even longer history. Moreover, when constellations are diversified in this manner, the development of innovative policies is likely. And, as curious as it may appear, such policies are even more likely when the constellation cannot forego a historical corporatist practice common in agriculture because it may offset the consequences of collective action problems, potentially severe on the environmental front. When distributing veto powers evenly within a relatively diversified constellation comprising actors from sectors affected by collective action problems, corporatist networks create good conditions for innovation since everyone's point of view needs to be addressed. Under these conditions, actors are required to come up with innovative solutions or welfare-enhancing bargains to accommodate every actor's preferred position. Certainly, such a corporatist network enabled the development of a comparatively promising environmental policy for agriculture in France. In short, as Pierson argues, past policy decisions have significant feedback effects, but more often than not they structure current policy development in a manner conducive to policy changes that depart from welfare state approaches.

Lastly, Pierson's proposals leave the impression that large state bureaucracies contribute significantly to perpetuating the status quo. These bureaucracies, he argues, employ experts with specific cognitive maps that filter new information, and they are accustomed to a limited range of policy solutions often implemented through administrative routines. Such an image confirms the popular perception that large bureaucracies have a negative effect on policy-making performance. In opposition to such a perception, the information presented in this book suggests that powerful bureaucracies, under certain conditions, might constitute a solution rather than a problem. Of course, as the main actor in a state-directed network, the USDA prevented taking advantage of the diversity contained in the American federal actor constellation. However, thanks to a stronger environment ministry, France was able to achieve a high level of environmental policy performance in the 1990s. At the present time, given the difficulty of organizing efficient environmental groups at the federal level in Canada, improvements in the balance between cohesion and diversity within the Ottawa constellation would demand a significant increase in the resources of the environmental bureaucracy. In Ontario, as the inquiry report on the Walkerton tragedy suggests, the environment ministry needs strengthening,[15] as does the agricultural ministry, whose autonomy vis-à-vis farm groups has recently been seriously undermined by the establishment of a clientelist policy network. In opposition to much of the literature, but in line with Weiss, I have attempted to illustrate the useful contribution of government bureaucracies to the governance of the essential interdependence between the state and civil society.

Internationalization and Regionalization

The internationalization and the regionalization of policy making provide a third theoretical source of apprehension for policy-making performance. On the one hand, when internationalization or regionalization adopts the form of intergovernmentalism, it creates a pattern of exclusion. Intergovernmental forums exclusively involve government executives and function in relative secrecy. Thus transporting policy decisions into these forums would allow state executives to forego the public debates democratic institutions tend to encourage. Such intergovernmental-stimulated autonomy, albeit accompanied by eroding state capacity in many sectors, would serve to disconnect policy makers from civil society, creating economy-oriented state-directed networks. The policies emerging out of such arrangements would mostly be desired and supported by a narrow coalition of actors working in the export sector.[16]

On the other hand, the European multilevel governance system, it has been argued, increases the fragmentation of the actor constellation. However, opinions diverge as to whether such a change represents a welcome break from the harmful exclusionary cohesion of member-state constellations or whether it transgresses the cohesion threshold necessary for efficient decision making.[17] Nevertheless, some authors have persuasively argued that European integration undermines the corporatist policy networks that have traditionally structured policy making at the member-state level. By eroding the power of member states, European integration has reduced the capacity of national policy makers to participate in corporatist bargains, while European institutional conditions prevent the organization of similar bargains at the supranational level.[18]

As observed in France, multilevel governance has increased the diversity of the agricultural actor constellation without threatening the cohesion necessary for efficient decision making. Responding to a novel institutional logic, the European Union can encourage policy progress blocked at the member-state level. Lagging behind that of other member states, environmental policy in France benefited from a European impulse. With the implementation of the Nitrate Directive, France had no choice but to reposition its environment ministry at the centre of the agro-environmental actor constellation. While the significance of this change in terms of performance cannot be denied, such a change corresponds to an adjustment in France's policy-making process rather than to a complete overhaul as implied by the multilevel governance metaphor. Once the Nitrate Directive was adopted, European actors played a relatively withdrawn role in France's environmental policy development. Commission officials, for example, are only periodically present and consequently are, at the very best, situated at the periphery of the French agro-environmental actor constellation. Therefore, such a constellation can hardly be characterized as a multilevel constellation.

Even more significantly, the involvement of the European Union in problems of agricultural pollution has not eroded the capacity of the French state to tackle the problem but rather has increased it. Under such circumstances, the corporatist policy network, which has presided over the development of agricultural policy in France, has not been threatened. Regional integration has not eroded the capacity of the French state to participate in a corporatist bargain.[19] As shown above, not only did corporatism survive the increased diversity of the actor constellation, but it contributed to France's high agro-environmental policy-making performance. As Scharpf points out, however, responses to European integration in terms of problem-solving capacity may vary from one sector to the next, with the environment, studies reveal, standing as one of the sectors where policy performances appear to increase.[20] Nevertheless, this book shows that associating overwhelmingly negative policy-making consequences with European integration can be misleading.

The structural effect of regional integration on actor constellations and networks, when such integration functions within the mode of intergovernmental relations, might also have been overstated. The state-directed network in the United States at the federal level had nothing to do with internationalization or regional integration. The strength of the USDA does not come from its capacity to transfer policy making to intergovernmental forums, and federal agro-environmental decisions were not meant to please a coalition of producers who export their commodities – although federal decisions possibly did not displease exporters as much as state-level decisions.

Policy decisions in Canada more often present traits likely to please exporters of agricultural commodities. Canadian policy makers often justify their aversion for command-and-control regulations in terms of international competitiveness. However, poor Canadian agro-environmental policy choices were made because of long-lasting serious deficiencies in the domestic actor constellations and policy networks – deficiencies that have little to do with the recent exposure of Canadian agriculture to internationalization and regionalization. Since the conclusion of the Uruguay Round of international trade negotiations and the adoption of NAFTA, alternatives to protect domestic agriculture are limited, Canadian policy makers claim, hence the importance of not overburdening farmers with stringent environmental obligations. Naturally, this reasoning would also apply to France and the United States, two countries that have nevertheless imposed intrusive regulations on farmers. While France and the United States are larger exporters of agricultural commodities than is Canada, agriculture occupies comparable shares of the economies of the three countries.[21] Consequently, helping farmers to meet environmental obligations would likely constitute no more of a burden on Canadian public finances than it already does in France or in the United States.

In any case, the agro-environmental actor constellations in Canada were alarmingly cohesive much before the conclusion of the Uruguay Round and NAFTA. And the cause of this cohesion is not that environmental policy making has been moved into closed intergovernmental forums but rather that environmental groups and environmental state bureaucracies have historically been weak in this country. In consequence, internationalization in Canada appears to simply take the form of a discourse on competitiveness as a means to externalize the blame for relatively ineluctable policy choices in the current Canadian institutional context. This analysis confirms Hirst's view that globalization, far from being a structural force undermining governance capacity, constitutes nothing more than a rhetoric used for political ends.[22] While it may be a good idea to remain vigilant in regard to the progress of internationalization and regionalization, this book certainly suggests that these phenomena have not caused an alarming overhaul of policy networks, at least not in the environmental sector.

Conclusion

Significant academic efforts have gone into documenting the declining legitimacy of public governance institutions.[23] Utilitarian principles being popular, this crisis of confidence is taken in several of these studies as evidence in itself that governance structures are inadequate.[24] Inputs and outputs, Scharpf argues, can serve to legitimize governance structures.[25] Inputs can legitimize governance structures when they encourage public participation, and outputs can legitimize them when results are consistent with the common good. In line with Scharpf, I argued in Chapter 1 that output-oriented legitimacy is crucial to modern states because accommodating all the demands of large and diversified populations occurs only under exceptional circumstances. If citizens cannot always expect to participate in policy making in a meaningful manner, they should at least expect governance structures to produce public policies enhancing the collective welfare. Citizens' dissatisfaction, utilitarians would have us believe, indicates that governance structures are incapable of producing such policies.

Precisely because utilitarian principles are popular, few academics have ventured into questioning the assumption of inadequate governance underlying the confidence crisis. Taking for granted the inadequacy of governance structures, policy studies, more often than not, feed into the confidence crisis by constructing theories that inspire nothing but distrust.

Meant to determine whether the public is right to distrust governance structures, this book closely examined some of these theories. Overall, I argue that the confidence crisis is not entirely justified. Governance structures, understood here as policy networks, can display a high potential for problem solving, even though some theories of agenda setting, welfare state politics, and globalization suggest otherwise. Although unequally distributed among countries, this high potential for problem solving was even found to be increasing in France. Even in the difficult political context of today, networks can be sufficiently diversified to enable the "power of joint action" and cohesive enough to function adequately. State actors can remain capable of leadership in networks establishing close connections with civil society actors possessing interdependent capacities.

Therefore, this book argues that theoretical constructions presenting policy networks as generally incapable of addressing collective problems are often misleading. When nourished by theories suggesting that policy processes are essentially random, that interest groups are too powerful, that state bureaucracies are unmanageable, and that policy capacities are eroded by internationalization, the confidence crisis, I suggest, is a matter of misplaced distrust. Distrust is misplaced, first and foremost, because governance structures, more often than not, deserve citizens' confidence and, second, because interest groups, state bureaucracies, and even regional integration, when it takes the form of multilevel governance, are essential elements of adequate policy-making performance.

Notes

Chapter 1: Introduction

1 Ronald Inglehart, *Modernization and Postmodernization: Cultural, Economic, and Political Change in 43 Societies* (Princeton: Princeton University Press, 1997), 79.
2 Robert D. Putnam, Susan J. Pharr, and Russell J. Dalton, "Introduction: What's Troubling the Trilateral Democracies?" in *Disaffected Democracies: What's Troubling the Trilateral Countries?* ed. Susan J. Pharr and Robert D. Putnam (Princeton: Princeton University Press, 2000), 27.
3 Maureen Mancuso et al., *A Question of Ethics: Canadians Speak Out* (Toronto: Oxford University Press, 1998).
4 François Ricard, *La génération lyrique* (Montreal: Boréal, 1992).
5 Inglehart, *Modernization and Postmodernization.*
6 Putnam, Pharr, and Dalton, "Introduction," 25.
7 Ann Larason Schneider and Helen Ingram, *Policy Design for Democracy* (Lawrence: University Press of Kansas, 1997), 4.
8 Jon Pierre and B. Guy Peters, *Governance, Politics and the State* (New York: St. Martin's Press, 2000), 1.
9 R.A.W. Rhodes, *Understanding Governance: Policy Networks, Governance, Reflexivity and Accountability* (Buckingham: Open University Press, 1997).
10 Susan Strange, *The Retreat of the State: The Diffusion of Power in the World Economy* (Cambridge: Cambridge University Press, 1996).
11 Schneider and Ingram make a similar argument. See *Policy Design for Democracy*, 8.
12 Fritz W. Scharpf, *Games Real Actors Play: Actor-Centered Institutionalism in Policy Research* (Boulder: Westview Press, 1997).
13 Fritz W. Scharpf, *Governing in Europe: Effective and Democratic?* (Oxford: Oxford University Press, 1999), 8.
14 This perspective is the one adopted by Schneider and Ingram, *Policy Design for Democracy*.
15 Scharpf, *Games Real Actors Play*, 155.
16 Ibid., 153.
17 Ibid.
18 Gregory E. McAvoy, *Controlling Technocracy: Citizen Rationality and the Nimby Syndrome* (Washington, DC: Georgetown University Press, 1999).
19 Rhodes, *Understanding Governance*, 9-10.
20 For a thorough criticism of this approach to networks, see Keith Dowding, "Model or Metaphor? A Critical Review of the Policy Network Approach," *Political Studies* 43 (1995): 136-58.
21 Hans Th.A. Bressers and Laurence J. O'Toole Jr., "The Selection of Policy Instruments: A Network-Based Perspective," *Journal of Public Policy* 18 (1998): 232.
22 William D. Coleman and Christine Chiasson, "State Power, Transformative Capacity and Adapting to Globalization: An Analysis of French Agricultural Policy 1960-2000," *Journal of European Public Policy* 9, 2 (2002): 168-85; Grace Skogstad, "Ideas, Paradigms and Institutions:

Agricultural Exceptionalism in the European Union and the United States," *Governance: An International Journal of Policy and Administration* 11, 4 (1998): 463-90; Éric Montpetit, "Europeanization and Domestic Politics: Europe and the Development of a French Environmental Policy for the Agricultural Sector," *Journal of European Public Policy* 7, 4 (2000): 576-92; David Marsh and Martin Smith, "Understanding Policy Networks: Towards a Dialectical Approach," *Political Studies* 48 (2000): 4-21.

23 William D. Coleman, "From Protected Development to Market Liberalism: Paradigm Change in Agriculture," *Journal of European Public Policy* 5, 4 (1998): 632-51.

24 Michael M. Atkinson and William D. Coleman, "Strong States and Weak States: Sectoral Policy Networks in Advanced Capitalist Economies," *British Journal of Political Science* 19 (1989): 47-67.

25 Linda Weiss, *The Myth of the Powerless State* (Ithaca: Cornell University Press, 1998).

26 Lord Kennet in 1969, quoted in Arnold J. Heidenheimer, Hugh Heclo, and Carolyn Teich Adams, *Comparative Public Policy: The Politics of Social Choice in America, Europe, and Japan,* 3rd ed. (New York: St. Martin's Press, 1990), 330.

27 Inglehart, *Modernization and Postmodernization.*

28 See John Mark Hansen, *Gaining Access: Congress and the Farm Lobby, 1919-1981* (Chicago: University of Chicago Press, 1991).

29 Wyn P. Grant, "The Limits of the Common Agricultural Policy Reform and the Option of Renationalisation," *Journal of European Public Policy* 2 (1995): 1-18.

30 Janice Gross Stein, *The Cult of Efficiency* (Toronto: Anansi, 2001).

31 Ibid., 155.

32 Putnam, Pharr, and Dalton, "Introduction," 22.

33 See Stein, *The Cult of Efficiency*, 154-84.

34 Robert D. Putnam, *Making Democracy Work: Civic Traditions in Modern Italy* (Princeton: Princeton University Press, 1993), 66.

35 Success in controlling soil erosion in several countries might in fact be less the result of government programs than the result of the development of cost-effective no-till technologies.

36 Putnam, *Making Democracy Work*, 65-66.

37 Peter M. Haas, "Introduction: Epistemic Communities and International Policy Coordination," *International Organization* 46, 1 (1992): 1-36.

38 William D. Coleman, "From Protected Development to Market Liberalism: Paradigm Change in Agriculture," *Journal of European Public Policy* 5, 4 (1998): 632-51; William D. Coleman, Grace D. Skogstad, and Michael M. Atkinson, "Paradigm Shifts and Policy Networks: Cumulative Change in Agriculture," *Journal of Public Policy* 16, 3 (1997): 273-301.

39 For data on the use of commercial fertilizers between 1950 and 1987, I used Food and Agriculture Organization of the United Nations, *The State of Food and Agriculture* (Rome: FAO, 1970, 1990). For 1994 data see Organization for Economic Cooperation and Development, *OECD Environmental Data: Compendium 1997* (Paris: OECD, 1997).

40 Fertilization is of course not the only factor that contributed to increased yields.

41 United States Department of Agriculture, *Agricultural Statistics 1997* (Washington, DC: United States Printing Office, 1997), I-8; United States Department of Agriculture, *Agricultural Statistics 1961* (Washington, DC: United States Printing Office, 1961), 6.

42 For statistics see Organization for Economic Cooperation and Development, *Economic Accounts for Agriculture* (Paris: OECD, published annually).

43 William D. Coleman, Michael M. Atkinson, and Éric Montpetit, "Against the Odds: Retrenchment in Agriculture in France and the United States," *World Politics* 49, 4 (July 1997): 453-81.

44 Pierre Rainelli and Dominic Vermersch, "Thematic Network on the CAP and the Environment in the European Union: A French Report" (unpublished paper, 1997). Competitive pressure also encourages farm concentration in the United States and Canada. See Owen J. Furuseth, "Restructuring of Hog Farming in North Carolina: Explosion and Implosion," *Professional Geographer* 49, 4 (1997): 391-403.

45 Mark Sproule-Jones, *Restoration of the Great Lakes: Promises, Practices, Performances* (Vancouver: UBC Press, 2002), 40-41.

46 Standing Committee on Agriculture, Fisheries and Forestry, *Soil at Risk* (Ottawa: Senate of Canada, 1984).

47 United States Environmental Protection Agency, "EPA to Better Protect Public Health and the Environment from Animal Feeding Operations" (headquarters press release, 5 March 1998).

48 Institut français de l'environnement, *Agriculture et environnement: Les indicateurs* (Paris: IFEN, 1997), quoted in *Le Monde,* 10 June 1997, 13.

49 Agriculture Canada, *Ontario Farm Groundwater Quality Survey* (Ottawa: Agriculture Canada, 1993). In Quebec, agricultural pollution is well documented because of a program called Réseau-Rivières. See Ministère de l'Environnement et de la Faune, *Le réseau-rivières: Un baromètre de la qualité de nos cours d'eau* (Quebec: MEF, 1991). See also David Berryman and Isabelle Giroux, *La contamination des cours d'eau par les pesticides dans les régions de culture intensive de maïs au Québec: Campagne d'échantillonnage de 1992 et 1993* (Quebec: MEF, 1994).

50 Food and Agriculture Organization of the United Nations, *Sustainability Issues in Agricultural and Rural Development Policies,* vol. 1, ed. F. Pétry (Rome: FAO, 1995), 2.19.

51 Here is a sample of a few newspaper articles that were published on the question: Karen Unland, "Pig Odour Raises a Stink in Quebec," *Globe and Mail,* 21 December 1996; Pat Stith, "The Smell of Money," *News and Observer,* 24 February 1995; Jean Le Doux, "Les bonnes recettes de deux producteurs bretons," *Ouest France,* 13 May 1997. For a broadly diffused study on the question of odours conducted in North Carolina by a group of researchers from North Carolina State University and Duke University, see Swine Odor Task Force, *Options for Managing Odor* (Raleigh: North Carolina Research Service, North Carolina State University, 1995).

52 For more information on the contribution of agriculture to a number of environmental problems, including biodiversity, soil erosion, and ozone depletion, in an easy-to-read format, see Food and Agriculture Organization of the United Nations, *Sustainability Issues in Agricultural and Rural Development Policies,* vols. 1 and 2.

53 Don Behm, "City Still on the Alert for Signs of Cryptosporidium," *Milwaukee Journal Sentinel,* 13 April 1998.

54 Douglas Powell, "Deadly Bacteria Show Up in Surprising Places," *Globe and Mail,* 25 January 1997.

55 Ken Killpatrick, "Concern Grows about Pollution from Megafarms," *Globe and Mail,* 30 May 2000.

56 Martine Valo, "L'État mis en cause pour son laxisme face à la pollution des eaux par les nitrates," *Le Monde,* 19 April 2001.

57 The chemical has since been banned by the European Union. On France see also Jean-Paul Dufour, "Quand l'ozone des champs envahit les villes," *Le Monde,* 8 September 1998.

58 Organization for Economic Cooperation and Development, *Environmental Performance Review: France* (Paris: OECD, 1997), 22.

59 Terry D. Garcia, "Testimony before the Subcommittee on Human Resources, Committee on Government Reform and Oversight, US House of Representatives" (25 September 1997).

60 See Pat Stith, Joby Warrick, and Melanie Sill, "Boss Hog: North Carolina's Pork Revolution," *News and Observer,* 12 May 1996, reprinted edition.

61 Putnam, Pharr, and Dalton, "Introduction," 21.

62 These statistics come from the FAOSTAT database, accessible from the Food and Agriculture Organization's website: <http://www.fao.org> (July 2002).

63 These statistics again come from the FAOSTAT database.

64 See Kathleen Thelen and Sven Steinmo, "Historical Institutionalism in Comparative Politics," in *Structuring Politics: Historical Institutionalism in Comparative Analysis,* ed. Sven Steinmo, Kathleen Thelen, and Frank Longstreth (New York: Cambridge University Press, 1992), 1-32.

65 Pierre Muller, *Le technocrate et le paysan* (Paris: Éditions ouvrières, 1984); John T.S. Keeler, *The Politics of Neocorporatism in France: Farmers, the State, and Agricultural Policy-Making in the Fifth Republic* (Oxford: Oxford University Press, 1987).

66 William P. Browne, *Cultivating Congress: Constituents, Issues, and Interests in Agricultural Policymaking* (Lawrence: University Press of Kansas, 1995).

67 Grace Skogstad, *The Politics of Agricultural Policy-Making in Canada* (Toronto: University of Toronto Press, 1987).
68 Weiss, *The Myth of the Powerless State.*

Chapter 2: Assessing Policy-Making Performance

1 Vincent Lemieux, *L'étude des politiques publiques: Les acteurs et leur pouvoir* (Sainte-Foy: Les Presses de l'Université Laval, 1995), Chapter 9.
2 To illustrate a similar point, Stein quotes the evaluations of Ontario's high schools conducted by the Fraser Institute. Janice Gross Stein, *The Cult of Efficiency* (Toronto: Anansi, 2001), 162.
3 Robert D. Putnam, *Making Democracy Work: Civic Traditions in Modern Italy* (Princeton: Princeton University Press, 1993), 63.
4 Fritz W. Scharpf, *Crisis and Choice in European Social Democracy* (Ithaca: Cornell University Press, 1987); Peter Hall, *Governing the Economy: The Politics of State Intervention in Britain and France* (New York: Oxford University Press, 1986).
5 Sven Steinmo, *Taxation and Democracy: Swedish, British, and American Approaches to Financing the Modern State* (New Haven: Yale University Press, 1993), 35-49.
6 Leslie A. Pal, *Beyond Policy Analysis: Public Issue Management in Turbulent Times* (Scarborough: Nelson, 1997), 7.
7 The OECD publishes environmental data in a yearly environmental data compendium.
8 For an elaborated discussion of the problem of comparing environmental performance, see David Vogel, *National Styles of Regulation: Environmental Policy in Great Britain and the United States* (Ithaca: Cornell University Press, 1986), 147-53.
9 John Detlef, "Environmental Performance and Policy Regimes," *Policy Science* 31 (1998): 107-31.
10 Vogel, *National Styles of Regulation*, 147.
11 A. Myrick Freeman III, "Economics, Incentives, and Environmental Regulation," in *Environmental Policy in the 1990s,* 3rd ed., ed. Norman J. Vig and Michael E. Kraft (Washington, DC: Congress Quarterly Press, 1997), 187-207.
12 For a summary of this literature, see Giandomenico Majone, *Evidence, Argument and Persuasion in the Policy Process* (New Haven: Yale University Press, 1989), Chapter 6.
13 Peter M. Haas, "Introduction: Epistemic Communities and International Policy Coordination," *International Organization* 46, 1 (1992): 1-36; Peter M. Haas, "Do Regimes Matter? Epistemic Communities and Mediterranean Pollution Control," *International Organization* 43, 3 (1989): 377-403.
14 Haas, "Introduction," 23; Peter M. Haas and Ernest B. Haas, "Learning to Learn: Improving International Governance," *Global Governance* 1 (1995): 260-62.
15 Majone, *Evidence, Argument, and Persuasion in the Policy Process.*
16 Haas, "Introduction," 4.
17 Ibid., 23.
18 Ibid., 27.
19 Haas, "Do Regimes Matter?" 401.
20 Ibid., 385.
21 Barry G. Rabe and Janet B. Zimmerman, "Beyond Environmental Regulatory Fragmentation," *Governance: An International Journal of Policy and Administration* 8, 1 (1995): 68-69.
22 Freeman III, "Economics, Incentives, and Environmental Regulation"; Chulho Jung, Kerry Krutilla, and Roy Boyd, "Incentives for Advanced Pollution Abatement Technology at the Industry Level: An Evaluation of Policy Alternatives," *Journal of Environmental Economics and Management* 30 (1996): 95-111; Richard Lotspeich, "Comparative Environmental Policy: Market-Type Instruments in Industrialized Capitalist Countries," *Policy Studies Journal* 26, 1 (1998): 85-104.
23 Majone, *Evidence, Argument and Persuasion in the Policy Process*, 141.
24 Grace Skogstad, "Ideas, Paradigms and Institutions: Agricultural Exceptionalism in the European Union and the United States," *Governance: An International Journal of Policy and Administration* 11, 4 (1998): 463-90.
25 Pierre Lascoumes, *L'éco-pouvoir: Environnement et politiques* (Paris: Édition la découverte, 1994).

26 C.R.W. Spedding, *An Introduction to Agricultural Systems* (London: Elsevier Applied Science, 1988), 37.
27 Spedding, *An Introduction to Agricultural Systems*, Chapter 1.
28 See J.L. Hatfield and D.L. Karlen, eds., *Sustainable Agriculture Systems* (Boca Raton: Lewis Publishers, 1994).
29 Joyti K. Parikh, "Introduction," in *Sustainable Development in Agriculture,* ed. Joyti K. Parikh (Dordrecht: Martimus Nijhoff Publishers, 1988), 2.
30 A typical statement: "Local research on the economic and physical performance of recommended practices can improve practice adoption." Marc O. Ribaudo, "Lessons Learned about the Performance of USDA Agricultural Nonpoint Source Pollution Programs," *Journal of Soil and Water Conservation* 53, 1 (1998): 8.
31 Walter N. Thurman, *Assessing the Environmental Impact of Farm Policies* (Washington, DC: AEI Press, 1995), 12-13.
32 For a general overview of the literature on market failures, see Charles Wolf Jr., "Carences du marché et carences hors-marché: Comparaison et évaluation," *Politiques et Management Public* 5, 1 (March 1987): 57-84; Charles Wolf Jr., *Market or Governments: Choosing between Imperfect Alternatives* (Cambridge: MIT Press, 1988).
33 Alfons Weersink et al., "Economic Instruments and Environmental Policy in Agriculture," *Canadian Public Policy* 24, 3 (1998): 312-13; Organization for Economic Cooperation and Development, *Environmental Performance in OECD Countries: Progress in the 1990s* (Paris: OECD, 1996), 22-23.
34 Pierre Rainelli, "Intensive Livestock Production in France and Its Effects on Water Quality in Brittany," in *Towards Sustainable Agricultural Development,* ed. M.D. Young (London: Belhaven Press, 1992), 126-44.
35 Haas, "Do Regimes Matter?" 386.
36 Ann Larason Schneider and Helen Ingram, *Policy Design for Democracy* (Lawrence: University Press of Kansas, 1997), 54; Michel Callon, Pierre Lascoumes, and Yannick Barthe, *Agir dans un monde incertain* (Paris: Seuil, 2001), Chapter 3.
37 See Groupe de Bruges, *Agriculture: Un tournant nécessaire* (Paris: Éditions de l'aube, 1996).
38 Haas and Haas, "Learning to Learn," 274, table.
39 In some countries, policy makers use science as a means to legitimize political decisions. In these instances, policy makers normally identify a compatible expert community and rely solely on that community for policy advice. Their ability to follow such a route, however, rests in part on institutional factors. Where the policy process is more adversarial and less dominated by strong state actors, various political interests are tempted to hire their own community of experts to legitimize their policy positions. Competing epistemic communities are naturally more likely to be active in these latter settings and thus science is unlikely to emerge as a clear guide to policy making. See Arnold J. Heidenheimer, Hugh Heclo, and Carolyn Teich Adams, *Comparative Public Policy: The Politics of Social Choice in America, Europe, and Japan,* 3rd ed. (New York: St. Martin's Press, 1990), 322-23.
40 Peter Bachrach and Morton S. Baratz, "Two Faces of Power," *American Political Science Review* 56 (1961): 947-52; Michael Howlett and M. Ramesh, *Studying Public Policy: Policy Cycles and Policy Subsystems* (Toronto: Oxford University Press, 1995), 4-5.
41 Schneider and Ingram, *Policy Design for Democracy*, 55.
42 Rabe and Zimmerman, "Beyond Environmental Regulatory Fragmentation."
43 Rainelli, "Intensive Livestock Production in France and Its Effects on Water Quality in Brittany," 139.
44 Fritz W. Scharpf, *Games Real Actors Play: Actor-Centered Institutionalism in Policy Research* (Boulder: Westview Press, 1997), 89-90.
45 Skogstad, "Ideas, Paradigms and Institutions."
46 Groupe de Bruges, *Agriculture: Un tournant nécessaire.*
47 Natacha Lajoie and François Blais, "Une réconciliation est-elle possible entre l'environnement et le marché? Une évaluation critique de deux tentatives," *Politique et Sociétés* 18, 3 (1999): 49-77.
48 John Kincaid, "Intergovernmental Costs and Coordination in U.S. Environmental

Protection," in *Federalism and the Environment: Environmental Policymaking in Austria, Canada, and the United States,* ed. Kenneth M. Holland and Frederick Lee Morton (Westport: Greenwood Press, 1996), 79-101; Kathryn Harrison, *Passing the Buck: Federalism and Canadian Environmental Policy* (Vancouver: UBC Press, 1996).

49 Lotspeich, "Comparative Environmental Policy," 99.

50 Weersink et al., "Economic Instruments and Environmental Policy in Agriculture," 321.

Chapter 3: Networks and Performance

1 Michael Howlett and M. Ramesh, *Studying Public Policy: Policy Cycles and Policy Subsystems* (Toronto: Oxford University Press, 1995), 105-7.

2 John W. Kingdon, *Agendas, Alternatives, and Public Policies* (New York: HarperCollins College Publishers, 1995). The original formulation belongs to Michael Cohen, James March, and Johan Olsen, "A Garbage Can Model of Organizational Choice," *Administrative Science Quarterly* 17 (March 1972): 1-25. See also Vincent Lemieux, *L'étude des politiques publiques: Les acteurs et leur pouvoir* (Sainte-Foy: Les Presses de l'Université Laval, 1995).

3 Kingdon, *Agendas, Alternatives, and Public Policies*, Chapter 6.

4 Ibid., Chapter 7.

5 Ibid., Chapter 5.

6 Michael Howlett, "Predictable and Unpredictable Windows: Institutional and Exogenous Correlates of Canadian Federal Agenda-Setting," *Canadian Journal of Political Science* 31, 3 (1998): 495-524.

7 On anticipation versus reaction, see Michael M. Atkinson and William D. Coleman, "Strong States and Weak States: Sectoral Policy Networks in Advanced Capitalist Economies," *British Journal of Political Science* 19 (1989): 47-67.

8 Howlett, "Predictable and Unpredictable Windows," 517.

9 William D. Coleman and Grace Skogstad, eds., *Policy Communities and Public Policy in Canada: A Structural Approach* (Mississauga: Copp Clark Pitman, 1990); Michael M. Atkinson and William D. Coleman, "Policy Networks, Policy Communities and the Problems of Governance," *Governance: An International Journal of Policy and Administration* 5, 2 (1992): 154-80; Frans van Waarden, "Dimensions and Types of Policy Networks," *European Journal of Political Research* 21 (1992): 29-52; Beate Kohler-Koch, "Catching up with Change: The Transformation of Governance in the European Union," *Journal of European Public Policy* 3, 3 (1996): 359-80; John Peterson, "States, Societies and the European Union," *West European Politics* 20, 4 (1997): 1-23; Lawrence Busch and Arunas Juska, "Beyond Political Economy: Actor Networks and the Globalization of Agriculture," *Review of International Political Economy* 4, 4 (1997): 688-708.

10 Coleman and Skogstad, eds., *Policy Communities and Public Policy in Canada.*

11 Fritz W. Scharpf, *Games Real Actors Play: Actor-Centered Institutionalism in Policy Research* (Boulder: Westview Press, 1997), 71.

12 Ibid., Chapter 6.

13 Ibid., 130-32.

14 Grace Skogstad, "Ideas, Paradigms and Institutions: Agricultural Exceptionalism in the European Union and the United States," *Governance: An International Journal of Policy and Administration* 11, 4 (1998): 463-90; William D. Coleman, "From Protected Development to Market Liberalism: Paradigm Change in Agriculture," *Journal of European Public Policy* 5, 4 (1998): 632-51; William D. Coleman, Grace Skogstad, and Michael M. Atkinson, "Paradigm Shifts and Policy Networks: Cumulative Change in Agriculture," *Journal of Public Policy* 16, 3 (1997): 273-301.

15 Scharpf, *Games Real Actors Play,* 131.

16 Thomas Risse, "Let's Argue! Communicative Action in World Politics," *International Organization* 54, 1 (2000): 1-39.

17 Scharpf, *Games Real Actors Play*, 126-30.

18 Barry G. Rabe and Janet B. Zimmerman, "Beyond Environmental Regulatory Fragmentation," *Governance: An International Journal of Policy and Administration* 8, 1 (1995): 69.

19 Anne Larason Schneider and Helen Ingram, *Policy Design for Democracy* (Lawrence: University Press of Kansas, 1997), Chapter 6.

20 Scharpf, *Games Real Actors Play*, 136.
21 George Tsebelis, "Decision Making in Political Systems: Veto Players in Presidentialism, Parliamentarism, Multicameralism, and Multipartism," *British Journal of Political Science* 25 (1995): 289-325.
22 Hank C. Jenkins-Smith and Paul A. Sabatier, "The Dynamics of Policy-Oriented Learning," in *Policy Change and Learning: An Advocacy Coalition Approach*, ed. Paul A. Sabatier and Hank C. Jenkins-Smith (Boulder: Westview Press, 1993), 41-56.
23 Scharpf, *Games Real Actors Play*, 98.
24 J. Rogers Hollingsworth and Robert Boyer, eds., *Contemporary Capitalism: The Embeddedness of Institutions* (Cambridge: Cambridge University Press, 1997); Peter B. Evans, *Embedded Autonomy: States and Industrial Transformation* (Princeton: Princeton University Press, 1995); Linda Weiss, *The Myth of the Powerless State* (Ithaca: Cornell University Press, 1998).
25 Ibid., Chapters 2 and 3.
26 Ibid., 38.
27 William D. Coleman, "From Protected Development to Market Liberalism: Paradigm Change in Agriculture," *Journal of European Public Po*licy 5, 4 (1998): 632-51; Wolfgang Streeck and Philippe C. Schmitter, "Community, Market, State – and Associations? The Prospective Contribution of Interest Governance to Social Order," in *Private Interest Government*, ed. Wolfgang Streeck and Philippe C. Schmitter (London: Sage Publications, 1985), 15-16.
28 Paul Pierson, "The New Politics of the Welfare State," *World Politics* 48 (1996): 143-79.
29 See Oliver E. Williamson, *Markets and Hierarchies: Analysis and Antitrust Implications* (New York: The Free Press, 1975).
30 Paul Pierson, *Dismantling the Welfare State? Reagan, Thatcher, and the Politics of Retrenchment* (Cambridge: Cambridge University Press, 1994), 44.
31 Paul Pierson, "Increasing Returns, Path Dependence, and the Study of Politics," *American Political Science Review* 94, 2 (2000): 251-67; Douglass C. North, *Institutions, Institutional Change and Economic Performance* (Cambridge: Cambridge University Press, 1990), 93-94.
32 Pierson, *Dismantling the Welfare State?* 44.
33 Pierson, "Increasing Returns, Path Dependence, and the Study of Politics."
34 Judith Goldstein, "The Impact of Ideas on Trade Policy: The Origins of U.S. Agricultural and Manufacturing Policies," *International Organization* 43, 1 (1989): 31-72; John Mark Hansen, *Gaining Access: Congress and the Farm Lobby, 1919-1981* (Chicago: University of Chicago Press, 1991).
35 Scharpf, *Games Real Actors Play*.
36 David Held et al., *Global Transformations: Politics, Economics and Culture* (Stanford: Stanford University Press, 1999), 16.
37 William D. Coleman and Anthony Perl, "Internationalized Policy Environments and Policy Network Analysis," *Political Studies* 47, 4 (1999): 701.
38 Gary Marks, Liesbet Hooghe, and Kermit Blank, "European Integration from the 1980s: State-Centric v. Multi-Level Governance," *Journal of Common Market Studies* 34, 3 (1996): 341-78; Thomas Risse-Kappen, "Exploring the Nature of the Beast: International Relations Theory and Comparative Policy Analysis Meet the European Union," *Journal of Common Market Studies* 34, 1 (1996): 53-80.
39 Richard Simeon, "Inside the Macdonald Commission," *Studies in Political Economy* 22 (1987): 167-79; Duncan Cameron and Daniel Drache, "Outside the Macdonald Commission: Reply to Richard Simeon," *Studies in Political Economy* 26 (1988): 173-80.
40 Jeffry A. Frieden and Ronald Rogowski, "The Impact of the International Economy on National Policies: An Analytical Overview," in *Internationalization and Domestic Politics*, ed. Robert O. Keohane and Helen V. Milner (Cambridge: Cambridge University Press, 1996), 25-47; Richard Clayton and Jonas Pontusson, "Welfare State Retrenchment Revisited: Entitlement Cuts, Public Sector Restructuring, and Inegalitarian Trends in Advanced Capitalist Societies," *World Politics* 51 (1998): 67-98; William D. Coleman, Michael M. Atkinson, and Éric Montpetit, "Against the Odds: Retrenchment in Agriculture in France and the United States," *World Politics* 49, 4 (July 1997): 453-81.
41 Robert D. Putnam, "Diplomacy and Domestic Politics," *International Organization* 42 (1988): 427-61.

42 William D. Coleman and Stefan Tangermann, "The 1992 CAP Reform, the Uruguay Round and the Commission: Conceptualizing Linked Policy Games," *Journal of Common Market Studies* 37, 3 (1999): 385-405. For a slightly different point of view, see Robert Paarlberg, "Agricultural Policy Reform and the Uruguay Round: Synergistic Linkage in a Two-Level Game?" *International Organization* 51, 3 (1997): 413-44.
43 Risse-Kappen, "Exploring the Nature of the Beast."
44 Coleman and Perl, "Internationalized Policy Environments and Policy Network Analysis," 703.
45 Kohler-Koch, "Catching up with Change."
46 Fritz W. Scharpf, *Governing in Europe: Effective and Democratic?* (Oxford: Oxford University Press, 1999).
47 Wolfgang Streeck, "From National Corporatism to Transnational Pluralism: European Interest Politics and the Single Market," in *Participation in Public Policy-Making: The Role of Trade Unions and Employers' Associations,* ed. Tiziano Treu (Berlin: Walter de Gruyter, 1992), 97-126. For a different but marginal point of view see Hugh Compston, "The End of National Policy Concertation? Western Europe Since the Single European Act," *Journal of European Public Policy* 5, 3 (1998): 507-26.
48 John Detlef, "Environmental Performance and Policy Regimes," *Policy Science* 31 (1998): 107-31.
49 Scharpf, *Governing in Europe*.

Chapter 4: France

1 Eve Fouilleux, "Réforme de la Pac, accord au Gatt: Quelles incidences sur les transferts financiers entre les Douze?" *Economie et Prévision* 117-18 (1995): 129-41.
2 John Ruggie, "International Regimes, Transactions, and Change: Embedded Liberalism in the Postwar Economic Order," *International Organization* 36 (1982): 379-415.
3 William D. Coleman, "From Protected Development to Market Liberalism: Paradigm Change in Agriculture," *Journal of European Public Policy* 5, 4 (1998): 632-51.
4 See also Pierre Rainelli, "Intensive Livestock Production in France and Its Effects on Water Quality in Brittany," in *Towards Sustainable Agricultural Development,* ed. M.D. Young (London: Belhaven Press, 1992), 126-44. Assemblée permanente des chambres d'agriculture, "Concentration très rapide de l'agriculture française depuis 1989," *Chambres d'Agriculture* 853 (March 1997): no pag. "Un tiers des ressources en eau potable sous la menace des nitrates," *Le Monde*, 10 June 1997, 13.
5 The Blair House Agreement effectively ended the dispute over the inclusion of agriculture in the world trade regime established by successive General Agreements on Tariffs and Trade (GATT).
6 William D. Coleman, Michael M. Atkinson, and Éric Montpetit, "Against the Odds: Retrenchment in Agriculture in France and the United States," *World Politics* 49, 4 (July 1997): 453-81.
7 See Eve Fouilleux, *La cogestion à la française à l'épreuve de l'Europe: L'exemple de la réforme de la Politique Agricole Commune* (Grenoble: Centre de recherche sur le politique, l'administration et le territoire, 1996).
8 Pierre Muller, *Le technocrate et le paysan* (Paris: Éditions ouvrières, 1984). On corporatism in France, see also Pierre Coulomb, "Les conférences annuelles entre corporatisme et populisme," in *Les agriculteurs et la politique,* ed. Pierre Coulomb et al. (Paris: Presses de la Fondation nationale des science politiques, 1990), 159-79.
9 Pierre Lascoumes, *L'éco-pouvoir: Environnement et politiques* (Paris: Édition la découverte, 1994).
10 To simplify things I have excluded government-sponsored research, which has increasingly sought to address agro-environmental problems.
11 Jan Douwe van der Ploeg and Gerrit van Dijk, eds., *Beyond Modernization: The Impact of Endogenous Rural Development* (Assen: Van Gorcum, 1995).
12 David Baldock and Philip Lowe, "The Development of European Agri-environmental Policy," in *The European Environment and CAP Reform: Policies and Prospects for Conservation,* ed. Martin Whitby (Wallingford: Cab International, 1996), 8-25.

13 Jean-Paul Billaud, "Article 19: Une gestion agricole au nom de l'environnement?" *Économie Rurale* 208/209 (1992): 137-41.
14 An official of the Ministry of Agriculture in Paris said that between 1989 and 1992 "the programs chosen were located in zones where the environmental problems were rather easy to solve such as le Marrais Poitevin, and les Vercores – that is, small sites where everybody including farmers could only admit that the environment had been significantly altered. It is true that the environmental impact of this first phase was nil; a few better maintained hectares represents nothing in relation to the total farm land in France." Interview, May 1997. See also Billaud, "Article 19."
15 The extensification of beef cattle production was achieved by compensating producers who agreed to reduce their herds or by compensating farmers who agreed to produce the same number of cattle but on expanded land.
16 Interview, May 1997.
17 Wyn P. Grant, "The Limits of the Common Agricultural Policy Reform and the Option of Renationalisation," *Journal of European Public Policy* 2 (1995): 3.
18 Farms are required to have a system to collect most of their waste into a three-month-capacity storage facility.
19 Lascoumes, *L'éco-pouvoir*, 106.
20 Loi du 9 janvier 1985 relative au développement et à la protection de la montagne and Loi du 8 janvier 1986 relative à l'aménagement, la protection et la mise en valeur du littoral.
21 Rainelli, "Intensive Livestock Production in France and Its Effects on Water Quality in Brittany," 134.
22 *L'Écho des Nitrates et des Phytos,* special Salon de l'Agriculture, 110 (February-March 1997).
23 See Richard Lotspeich, "Comparative Environmental Policy: Market-Type Instruments in Industrialized Capitalist Countries," *Policy Studies Journal* 26, 1 (1998): 89-90.
24 The general regulation comprises construction and manure application setbacks from neighbours and water bodies. Farms subject to the declaration regime are those with 50 to 200 beef cattle, 40 to 80 dairy cows, 50 to 450 pigs, or 5,000 to 20,000 poultry.
25 Farms subject to the authorization regime are those with over 200 beef cattle, over 80 dairy cows, over 450 pigs, and over 20,000 poultry.
26 Loi du 8 janvier 1993 sur la protection et la mise en valeur des paysages.
27 Anny Rousso, "Le droit du paysage: Un nouveau droit pour une nouvelle politique," *Courrier de l'Environnement de l'INRA* 26 (December 1995), <http://www.inra.fr/Internet/Produits/dpenv/roussc26.htm> (February 2003). A guidance law for land use planning was also adopted in 1995. However, like the landscape policy, it does not constitute a major departure from the past as far as agriculture is concerned. Arguably, the same could be said of the 1995 Loi sur le renforcement de la protection de l'environnement.
28 Patrice Cahart et al., *Rapport d'évaluation sur la gestion et le bilan du programme de maîtrise des pollutions d'origine agricole* (Paris: Ministère de l'Économie, des Finances et de l'Industrie; Ministère de l'Aménagement du territoire et de l'Environnement; Ministère de l'Agriculture et de la Pêche, 1999), 9-10.
29 "Plan de développement durable: Le coup d'envoi est donné," *Bima* 1449 (1996), <http://www.agriculture.gouv.fr/mapa"agriweb/bima/1449/49dv1.stm> (May 1998).
30 Michel Ledru, for the Conseil économique et social, "L'espace rural entre protection et contraintes," *Journal Officiel de la République Française* (Paris: Avis et rapports du Conseil économique et social, 1994), 118.
31 Coleman, Atkinson, and Montpetit, "Against the Odds."
32 European Commission, *CAP Working Notes: Agriculture and the Environment,* special issue (Brussels: European Commission, 1997), 24-25.
33 This program is similar to the previous extensification measures. Reforestation has become an accompanying measure of the CAP in its own right.
34 The term regional operation is not officially used. I use it following Jean-Marie Boisson and Henry Buller, "France," in *The European Environment and CAP Reform,* ed. Martin Whitby (Wallingford: Cab International, 1996), 124.
35 Interview, May 1997.
36 "Les mesures agri-environnement," *Entraid'Ouest* supplement (June 1997): 5.

37 These programs are far more complex than my discussion may imply. They touch particularly on aspects that are not relevant to agriculture or the environment. A full understanding of these instruments is far beyond the scope of this book. I nonetheless integrated them into my analysis, if only partially, in an effort to include all the policy instruments that fit the definition of agro-environmental policy instruments I have adopted.

38 *The Cork Declaration: A Living Countryside* (November 1996), <http://europa.eu.int/en/comm/dg06/new/cork.htm> (May 1998).

39 At <http://europa.eu.int/en/comm/dg06/rur/region.htm> (May 1998). The FNSEA also manages the Fonds de développement rural.

40 Boisson and Buller, "France," 130.

41 Arguing that policy has changed is different from arguing that the new regulations, the PMPOA, and other novel programs were successful or will be successful in changing agricultural practices. Observers often note implementation problems and lapses in compliance. An official with a *chambre d'agriculture* said: "The PMPOA is an eternal problem of implementation. We are done responding to one problem, and another one comes up right away." On compliance, the same interviewee said that "there are roughly 2,000 hog producers who played stupid in Brittany. That is, they 'forgot' that they needed to declare or ask for authorization to expand" (interview, June 1997). On the PMPOA, another interviewee said that "farmers take advantage of the program to undertake work on their farm that allows them to expand and thereby increase their potential for pollution in the long term" (interview, May 1997).

42 *Rapport de la France à la Commission du développement durable des Nations unies* (Paris: Ministère de l'Aménagement du territoire et de l'Environnement et Ministère des Affaires étrangères, 2000), 52-53; William D. Coleman and Christine Chiasson, "State Power, Transformative Capacity and Adapting to Globalization: An Analysis of French Agricultural Policy, 1960-2000," *Journal of European Public Policy* 9, 2 (2002): 180.

43 John Peterson, "Policy Networks and European Union Policy Making: A Reply to Kassim," *West European Politics* 18, 1 (1995): 389-407; John Peterson, "States, Societies and the European Union," *West European Politics* 20, 4 (1997): 1-23; Beate Kohler-Koch, "Catching up with Change: The Transformation of Governance in the European Union," *Journal of European Public Policy* 3, 3 (1996): 359-80; William D. Coleman and Anthony Perl, "Internationalizing Policy Environments and Policy Network Analysis," *Political Studies* 47, 4 (1999): 691-709; Maria Green Cowles, James Caporaso, and Thomas Risse, eds., *Transforming Europe: Europeanization and Domestic Change* (Ithaca: Cornell University Press, 2001).

44 Gary Marks, Liesbet Hooghe, and Kermit Blank, "European Integration from the 1980s: State-Centric v. Multi-Level Governance," *Journal of Common Market Studies* 34, 3 (1996): 341-78; Thomas Risse-Kappen, "Exploring the Nature of the Beast: International Relations Theory and Comparative Policy Analysis Meet the European Union," *Journal of Common Market Studies* 34, 1 (1996): 53-80.

45 École nationale d'administration, under the direction of Lucien Chabasson, *L'aménagement de l'espace rural,* vol. 2 (Paris: ENA, 1994), 672.

46 Lascoumes, *L'éco-pouvoir*, 193.

47 Interview, May 1997.

48 On this latter point see Boisson and Buller, "France," 130.

49 Interview, June 1997.

50 Bernard Kaczmarek, "La politique communautaire de l'eau," *Aménagement et Nature* 124 (March 1997): 25-36.

51 Interview, May 1997.

52 Blair House I was the first GATT agreement on the liberalization of agriculture. It was followed a few months later by Blair House II, which revisited some sections of Blair House I deemed too liberalizing by several farm groups.

53 See Fouilleux, "Réforme de la Pac, accord au Gatt," 129-41.

54 Interview, May 1997. A *jacquerie* is a violent protest.

55 Coleman and Chiasson, "State Power, Transformative Capacity and Adapting to Globalization," 179.

56 Louis Le Pensec, one of the agriculture ministers during the Jospin government, hired

Bertrand Hervieu to advise him. Hervieu, an academic, is also a member of the Groupe de Bruges, a group associated with the alternative agriculture epistemic community. See Groupe de Bruges, *Agriculture: Un tournant nécessaire* (Paris: Éditions de l'aube, 1996).

57 Muller, *Le technocrate et le paysan*; John T.S. Keeler, *The Politics of Neocorporatism in France: Farmers, the State, and Agricultural Policy-making in the Fifth Republic* (Oxford: Oxford University Press, 1987).
58 Fouilleux, *La cogestion à la française à l'épreuve de l'Europe*, 89.
59 Interview, May 1997.
60 Interview, June 1997.
61 My translation of Jean Salmon, "Environnement et élevage," *L'Information Agricole* 683 (November 1995): 17.
62 Scharpf, *Games Real Actors Play*, 127.
63 Cahart et al., *Rapport d'évaluation sur la gestion et le bilan du programme de maîtrise des pollutions d'origine agricole.*
64 Scharpf, *Games Real Actors Play*, 130-32.

Chapter 5: The United States

1 John W. Kingdon, *Agendas, Alternatives, and Public Policies* (New York: HarperCollins, 1995).
2 William P. Browne, *Cultivating Congress: Constituents, Issues, and Interests in Agricultural Policymaking* (Lawrence: University Press of Kansas, 1995).
3 See John Mark Hansen, *Gaining Access: Congress and the Farm Lobby, 1919-1981* (Chicago: University of Chicago Press, 1991); William P. Browne, *Private Interests, Public Policy, and American Agriculture* (Lawrence: University Press of Kansas, 1988).
4 David Vogel, *National Styles of Regulation: Environmental Policy in Great Britain and the United States* (Ithaca: Cornell University Press, 1986).
5 Kenneth Finegold and Theda Skocpol, *State and Party in America's New Deal* (Madison: University of Wisconsin Press, 1995).
6 United States Environmental Protection Agency, *Strategy for Addressing Environmental and Public Health Impacts from Animal Feeding Operations* (Washington DC: EPA, March 1998), 6.
7 Animal units correspond to 1,000 feeder cattle, 700 dairy cows, 2,500 pigs weighing over 25 kilograms, 100,000 laying hens or broilers with a continuous flow watering system. See United States Environmental Protection Agency, *Guide Manual on NPDES Regulations for Concentrated Animal Feeding Operations* (Washington, DC: EPA, December 1995), 6.
8 United States Environmental Protection Agency, *Clean Water Action Plan: Restoring and Protecting America's Waters* (Washington, DC: EPA, February 1998), 62.
9 United States Environmental Protection Agency, *Strategy for Addressing Environmental and Public Health Impacts from Animal Feeding Operations* (Washington, DC: EPA, March 1999).
10 United States Environmental Protection Agency, *Proposed Revisions to CAFO Regulations* (Washington, DC: EPA, 12 January 2001), 4 and 34.
11 Interview, September 1995.
12 See HR 1138 of the 107th Congress: A bill to amend section 402 of the Federal Water Pollution Control Act to provide that no permit shall be required for animal feeding operations within the boundaries of a State if the State has established and is implementing a nutrient management program for those animal feeding operations.
13 See Robert J. Morgan, *Governing Soil Conservation: Thirty Years of the New Decentralization* (Baltimore: Johns Hopkins Press, 1965).
14 Walter N. Thurman, *Assessing the Environmental Impact of Farm Policies* (Washington, DC: AEI Press, 1995), 45.
15 Willard W. Cochrane, *The Development of American Agriculture: A Historical Analysis* (Minneapolis: University of Minnesota Press, 1993), 154.
16 R. Douglas Hurt, *American Agriculture: A Brief History* (Ames: Iowa State University Press, 1994), 333.
17 The Wetland Reserve Program and the Water Bank are two programs related to the CRP but of minor importance. See Thurman, *Assessing the Environmental Impact of Farm Policies*, 44.
18 Ibid., 45-48.

19 In 1986, the first year of the CRP, total CRP spending was US$137,305,000. In 1994 it reached US$202,992,000. In 1995 it dropped back to its 1986 level. Source: United States Department of Agriculture, *Agricultural Statistics 1997* (Washington, DC: United States Printing Office, 1997), XII-8.
20 Thurman, *Assessing the Environmental Impact of Farm Policies*, 54-56.
21 Cochrane, *The Development of American Agriculture*, 254.
22 Alan L. Sutton and Don D. Jones, "Animal Waste Management Regulations: A Look into the Future" (unpublished manuscript, 1997).
23 Ibid.
24 The deadline was not met; however, a large number of operations now have a permit, and the delivery of permits continues.
25 See Blue Ribbon Study Commission on Agricultural Waste, *Report to the 1995 General Assembly of North Carolina, Regular Session* (Raleigh: May 1996); Jim Cummings, "Legislation Relating to Agricultural Waste and Regulations and Penalties on Pollution of Water Resources" (paper presented at the Agribusiness Law Conference, Campbell University, January 1997). See also House Bill 515 of the 1997 Session of the General Assembly of North Carolina.
26 In 1998 the state nevertheless imposed a three-year moratorium on the construction of earthen lagoons. On the various regulations, see Department of Natural Resources, "Confinement Feeding Operations," <http://www.state.ia.us/government/dnr/organiza/epd/wastewtr/feedlot/feedlt.htm> (May 1998). See also Environmental Protection Commission, "Adopted Rule Awaiting Publication: Chapter 65 Animal Feeding Operations" (Des Moines: Government of Iowa, 1997).
27 For information on Oklahoma regulations, see Oklahoma Department of Agriculture, "Concentrated Animal Feeding Operations" (news release, 17 December 1997), <http://www.oklaosf.state.ok.us/~okag/piarc/caforeq.html> (May 1998); Oklahoma House of Representatives, "Hog Regulation Bill Becomes Law" (news release, 4 June 1997), <http://www.lsb.state.ok.us/house/news407.htm> (May 1998); Oklahoma House of Representatives, "State House Divides Over Tougher Regulation of Hog Farms" (news release, 3 June 1998), <http://www.lsb.state.ok.us/house/news863.htm> (June 1998); Animal Waste and Water Quality Task Force, *Final Report* (Oklahoma City: Office of the Secretary of the Environment, December 1997).
28 Oklahoma House of Representatives, "Legislator Calls for Cooperative Control of Water Quality in Oklahoma" (news release, 13 May 1998), <http://www.lsb.state.ok.us/house/news808.htm> (June 1998).
29 Blue Ribbon Study Commission on Agricultural Waste, *Report to the 1995 General Assembly of North Carolina, Regular Session*, 3.
30 Cummings, "Legislation Relating to Agricultural Waste and Regulations and Penalties on Pollution of Water Resources." 7.
31 Browne, *Private Interests, Public Policy, and American Agriculture*.
32 Paul A. Sabatier, "Policy Change over a Decade or More," in *Policy Change and Learning: An Advocacy Coalition Approach*, ed. Paul A. Sabatier and Hank C. Jenkins-Smith (Boulder: Westview Press, 1993), 13-40.
33 Kingdon, *Agendas, Alternatives, and Public Policies*.
34 Hank C. Jenkins-Smith and Paul A. Sabatier, "The Dynamics of Policy-Oriented Learning," in *Policy Change and Learning: An Advocacy Coalition Approach*, ed. Paul A. Sabatier and Hank C. Jenkins-Smith (Boulder: Westview Press, 1993), 42.
35 Ibid., 48.
36 Interview, May 1998.
37 National Environmental Dialogue on Pork Production, *Comprehensive Environmental Framework for Pork Production Operations* (Washington, DC: America's Clean Water Foundation, December 1997).
38 It is important to note that environmental groups in the United States have better resources for exercising policy influence than do their French counterparts.
39 National Environmental Dialogue on Pork Production, *Comprehensive Environmental Framework for Pork Production Operations*, 2. Emphasis added.

40 Interview, May 1998.
41 Browne, *Cultivating Congress.*
42 See Morgan, *Governing Soil Conservation.*
43 Interview, May 1998.
44 Ibid.
45 See Morgan, *Governing Soil Conservation.*
46 See Barry G. Rabe, "Power to the States: The Promise and Pitfalls of Decentralization," in *Environmental Policy in the 1990s,* 3rd ed., ed. Norman J. Vig and Michael E. Kraft (Washington, DC: Congressional Quarterly Press, 1997), 31-52.
47 Finegold and Skocpol, *State and Party in America's New Deal.*
48 Interview, May 1998.
49 See Sabatier, "Policy Change over a Decade or More," 14; Barry G. Rabe, "Federalism and Entrepreneurship: Explaining American and Canadian Innovation in Pollution Prevention and Regulatory Integration," *Policy Studies Journal* 27 (1999): 288-306.
50 Interview, Raleigh, March 1998.

Chapter 6: Canada

1 Federal and Provincial Ministers of Agriculture, *National Environment Strategy for Agriculture and Agri-Food* (Ottawa: Agriculture and Agri-Food Canada, 1995), 1.
2 Kathryn Harrison, *Passing the Buck: Federalism and Canadian Environmental Policy* (Vancouver: UBC Press, 1996).
3 Éric Montpetit, "Sound Science and Moral Suasion, Not Regulation: Facing Difficult Decisions on Agricultural Non-Point Source Pollution," in *Canadian Environmental Policy: Context and Cases,* 2nd ed., ed. Debora Van Nijnatten and Robert Boardman (Don Mills: Oxford University Press, 2002), 274-85.
4 In addition to note 1, see Agriculture Canada, *Growing Together* (Ottawa: Agriculture Canada, 1989).
5 Environment Canada, *A Guide to the New Canadian Environmental Protection Act* (Ottawa: Minister of Public Works and Government Services, 2000), 13.
6 The Pesticides Act is similar.
7 Federal/Provincial/Territorial Advisory Committee on Canada's National Program of Action for the Protection of the Marine Environment from Land-based Activities, *Canada's National Program of Action for the Protection of the Marine Environment from Land-based Activities* (Ottawa: Minister of Public Works and Government Services Canada, 2000), 7.
8 The goal of the St. Lawrence Action Plan Vision 2000, Phase 3, is "to educate the key players in agriculture about environmental problems."
9 Agriculture and Agri-Food Canada, *Environmental Sustainability of Agriculture: Report of the Agri-Environmental Indicator Project* (Ottawa: Minister of Public Works and Government Services Canada, 2000).
10 Agriculture and Agri-Food Canada, *CARD: Fact Sheet* (Ottawa: Agriculture and Agri-Food Canada), at <http://www.agr.ca/policy/adapt/information/cardfactsheet.html> (July 2002).
11 <http://www.agr.ca/policy/adapt/adaptation_programs/adaptation_programs.html> (July 2002).
12 Environment Canada, *A Guide to the New Canadian Environmental Protection Act,* 5.
13 Jean Nolet, *Étude comparative de différentes réglementations concernant les nuisances en agriculture* (Quebec: Rapport remis au ministère de l'Environnement et de la Faune, 1996), 21.
14 Éric Montpetit, "Corporatisme québécois et performance des gouvernants: Analyse comparative des politiques environnementales en agriculture," *Politique et Sociétés* 18, 3 (1999): 79-98.
15 In the act, "normal farm practice" is defined as "a practice that is conducted in a manner consistent with proper and accepted customs and standards as established and followed by similar agricultural operations under similar circumstances and includes the use of innovative technology used with advanced management practices." Farm Practices Protection Act, Article 1.
16 When OMAFRA began a series of consultations on amendments to the Farm Practices Protection Act in 1997, there was no indication of major changes in the orientation of the act.

17 Under the Ontario Building Code, municipalities have to deliver building permits for any structure larger than 10 square meters, including manure storage facilities. When the structure is a farm building, the municipalities often require a project assessment by an agricultural engineer. OMAFRA's regional offices can provide this service. In certain cases, the Conservation Authorities have an advisory role for the issuing of municipal building permits. It should be noted that municipalities in Quebec have a similar responsibility in delivering building permits for agricultural projects, yet the project has been certified only once by the Ministère de l'Environnement et de la Faune (MEF).

18 AgCare, "New Chair Named to Environmental Farm Plan Working Group," *Update* 9, 3 (Summer 1999): n. pag.

19 Dennis R. O'Connor, *Report of the Walkerton Inquiry: The Events of May 2000 and Related Issues* (Toronto: Publication Ontario, 2002). Dennis R. O'Connor, *Report of the Walkerton Inquiry: A Strategy for Safe Drinking Water* (Toronto: Publication Ontario, 2002).

20 Secrétariat du Conseil du trésor, *Gérer la réglementation au Canada* (Ottawa: Ministre des Approvisionnements et Services, 1996).

21 Donald Lemaire, "The Stick: Regulation as a Tool of Government," in *Carrots, Sticks and Sermons: Policy Instruments and Their Evaluation,* ed. Marie-Louise Bemelmans-Videc, Ray C. Rist, and Evert Vedung (New Brunswick: Transaction Publishers, 1998), 66-67.

22 Standing Committee on Agriculture, Fisheries, and Forestry, *Soil at Risk* (Ottawa: Senate of Canada, 1984).

23 The Soil and Water Environmental Enhancement Program (SWEEP) and the National Soil Conservation Program.

24 Paul Pierson, "When Effect Becomes Cause: Policy Feedback and Political Change," *World Politics* 45 (1993): 595-628.

25 William D. Coleman, Grace D. Skogstad, and Michael M. Atkinson, "Paradigm Shifts and Policy Networks: Cumulative Change in Agriculture," *Journal of Public Policy* 16, 3 (1997): 287-88; Grace Skogstad, "Agricultural Policy," in *Border Crossings: The Internationalization of Canadian Public Policy,* ed. G. Bruce Doern, Leslie A. Pal, and Brian W. Tomlin (Toronto: Oxford University Press, 1996), 143-64.

26 Fritz W. Scharpf, *Games Real Actors Play: Actor-Centered Institutionalism in Policy Research* (Boulder: Westview Press, 1997), 130-32.

27 Linda Weiss, *The Myth of the Powerless State* (Ithaca: Cornell University Press, 1998).

28 More often than not, environmental groups have very limited resources. In Ontario, a large share of these resources is devoted to building an expertise in the areas of nuclear energy, toxic substances, air pollution, and industrial pollution, problems for which environmental groups are often consulted by the OMEE. Thus when an environment official in Ontario was asked whether environmental groups address agricultural pollution, he replied: "Not that I am aware of. Among the briefings and letters to ministers that I have seen in the past year, environmental groups are absent" (interview, 12 February 1997). See also Mark Winfield, "The Ultimate Horizontal Issue: The Environmental Policy Experiences of Alberta and Ontario, 1971-1993," *Canadian Journal of Political Science* 27 (1994): 129-52.

29 Standing Committee on Agriculture, Fisheries, and Forestry, *Soil at Risk.*

30 Interview, 12 February 1997.

31 Skogstad, "Farm Policy Community," 69.

32 Ibid., 78.

33 Interview, 13 March 1997.

34 Ibid.

35 Ontario Farm Environmental Coalition, *Our Farm Environmental Agenda* (Ontario: OFEC, 1992).

36 Scharpf elaborates on the different capacities for collective action of coalitions and associations in *Games Real Actors Play,* 54-56.

37 Farmers are permitted to request a refund from the organization to which they choose to pay dues. They have to make the request within ninety days of the deadline for paying the annual dues. Otherwise, all Ontario farmers are required by law to belong to one of the accredited general farm organizations.

38 The activities of the coalition are well described on the AgCare website: <http://www.agcare.org> (July 2002).

39 Skogstad argues that the general strength of OMAFRA is difficult to characterize, implying that it possesses weaknesses in some areas. The environment is definitely not one of its strong areas ("The Farm Policy Community," 78).

40 Weiss, *The Myth of the Powerless State.*

41 Ministry of Agriculture and Food, "Nutrient Management Act" (news release, 28 June 2002).

42 Éric Montpetit and William D. Coleman, "Policy Communities and Policy Divergence in Canada: Agro-Environmental Policy Development in Quebec and Ontario," *Canadian Journal of Political Science* 32, 4 (1999): 691-714.

43 G. Bruce Doern and Thomas Conway, *The Greening of Canada: Federal Institutions and Decision* (Toronto: University of Toronto Press, 1994), 82.

Chapter 7: Misplaced Distrust

1 Standing Committee on Agriculture, Fisheries, and Forestry, *Soil at Risk* (Ottawa: Senate of Canada, 1984).

2 Fritz W. Scharpf, *Games Real Actors Play: Actor-Centered Institutionalism in Policy Research* (Boulder: Westview Press, 1997).

3 Linda Weiss, *The Myth of the Powerless State* (Ithaca: Cornell University Press, 1998).

4 William D. Coleman, "Policy Communities and Policy Networks: Some Issues of Method" (paper prepared for presentation to the System of Government Conference, University of Pittsburgh, 1 November 1997).

5 David Vogel, "Representing Diffuse Interests in Environmental Policymaking," in *Do Institutions Matter? Government Capabilities in the United States and Abroad,* ed. R. Kent Weaver and Bert A. Rockman (Washington, DC: The Brookings Institution, 1993), 270; Éric Montpetit, "Sound Science and Moral Suasion, Not Regulation: Facing Difficult Decisions on Agricultural Non-Point Source Pollution," in *Canadian Environmental Policy: Context and Cases,* 2nd ed., ed. Debora Van Nijnatten and Robert Boardman (Don Mills: Oxford University Press, 2002), 274-85.

6 John Detlef, "Environmental Performance and Policy Regimes," *Policy Sciences* 31 (1998): 107-31.

7 As indicated in Chapter 6, I have explained Quebec's exceptional situation elsewhere: Éric Montpetit, "Corporatisme québécois et performance des gouvernants: Analyse comparative des politiques environnementales en agriculture," *Politique et Sociétés* 18, 3 (1999): 79-98.

8 Dennis R. O'Connor, *Report of the Walkerton Inquiry: The Events of May 2000 and Related Issues* (Toronto: Publication Ontario, 2002). Dennis R. O'Connor, *Report of the Walkerton Inquiry: A Strategy for Safe Drinking Water* (Toronto: Publication Ontario, 2002).

9 John W. Kingdon, *Agendas, Alternatives, and Public Policies* (New York: HarperCollins College Publishers, 1995).

10 Michael Howlett, "Predictable and Unpredictable Windows: Institutional and Exogenous Correlates of Canadian Federal Agenda-Setting," *Canadian Journal of Political Science* 31, 3 (1998): 495-524.

11 Pierre Muller, "Gouvernance européenne et globalisation," *Revue Internationale de Politique Comparée* 6 (1999): 707-17; Peter M. Haas, "Introduction: Epistemic Communities and International Policy Coordination," *International Organization* 46, 1 (1992): 1-36.

12 Paul Pierson, "The New Politics of the Welfare State," *World Politics* 48 (1996): 143-79.

13 Paul Pierson, "Increasing Returns, Path Dependence, and the Study of Politics," *American Political Science Review* 94, 2 (2000): 251-67.

14 William D. Coleman, Michael M. Atkinson, and Éric Montpetit, "Against the Odds: Retrenchment in Agriculture in France and the United States," *World Politics* 49, 4 (July 1997): 453-81.

15 O'Connor, *Report of the Walkerton Inquiry.*

16 Richard Clayton and Jonas Pontusson, "Welfare State Retrenchment Revisited: Entitlement Cuts, Public Sector Restructuring, and Inegalitarian Trends in Advanced Capitalist Societies," *World Politics* 51 (1998): 96-97.

17 Beate Kohler-Koch, "Catching up with Change: The Transformation of Governance in the European Union," *Journal of European Public Policy* 3, 3 (1996): 359-80; *versus* Fritz W. Scharpf, *Governing in Europe: Effective and Democratic?* (Oxford: Oxford University Press, 1999).
18 Wolfgang Streeck, "From National Corporatism to Transnational Pluralism: European Interest Politics and the Single Market," in *Participation in Public Policy-Making: The Role of Trade Unions and Employers' Associations,* ed. Tiziano Treu (Berlin: Walter de Gruyter, 1992), 97-126.
19 Hugh Compston makes similar observations in other sectors. See "The End of National Policy Concertation? Western Europe Since the Single European Act," *Journal of European Public Policy* 5, 3 (1998): 507-26.
20 Scharpf, *Governing in Europe*, 117.
21 See the FAO statistics presented on page 16 of Chapter 1.
22 Paul Hirst, "The Global Economy: Myths and Realities," *International Affairs* 73 (1997): 409-25.
23 For example: Ronald Inglehart, *Modernization and Postmodernization: Cultural, Economic, and Political Change in 43 Societies* (Princeton: Princeton University Press, 1997).
24 For example, several of the studies in Susan J. Pharr and Robert D. Putnam, eds., *Disaffected Democracies: What's Troubling the Trilateral Countries?* (Princeton: Princeton University Press, 2000).
25 Scharpf, *Games Real Actors Play*.

Bibliography

AgCare. "New Chair Named to Environmental Farm Plan Working Group." *Update* 9, 3 (Summer 1999).

Agriculture and Agri-Food Canada. *Environmental Sustainability of Agriculture: Report of the Agri-Environmental Indicator Project.* Ottawa: Minister of Public Works and Government Services Canada, 2000.

Agriculture Canada. *Growing Together.* Ottawa: Agriculture Canada, 1989.

–. *Ontario Farm Groundwater Quality Survey.* Ottawa: Agriculture Canada, 1993.

Animal Waste and Water Quality Task Force. *Final Report.* Oklahoma City: Office of the Secretary of the Environment, December 1997.

Assemblée permanente des chambres d'agriculture. "Concentration très rapide de l'agriculture française depuis 1989." *Chambres d'Agriculture* 853 (March 1997): no pag.

Atkinson, M., and W.D. Coleman. "Strong States and Weak States: Sectoral Policy Networks in Advanced Capitalist Economies." *British Journal of Political Science* 19 (1989): 47-67.

–. "Policy Networks, Policy Communities and the Problems of Governance." *Governance: An International Journal of Policy and Administration* 5, 2 (1992): 154-80.

Bachrach, P., and M.S. Baratz. "Two Faces of Power." *American Political Science Review* 56 (1961): 947-52.

Baldock, D., and P. Lowe. "The Development of European Agri-environmental Policy." In *The European Environment and CAP Reform: Policies and Prospects for Conservation,* edited by M. Whitby, 8-25. Wallingford: Cab International, 1996.

Behm, D. "City Still on the Alert for Signs of Cryptosporidium." *Milwaukee Journal Sentinel,* 13 April 1998.

Berryman, D., and I. Giroux. *La contamination des cours d'eau par les pesticides dans les régions de culture intensive de maïs au Québec: Campagne d'échantillonnage de 1992 et 1993.* Quebec: MEF, 1994.

Billaud, J.-P. "Article 19: Une gestion agricole au nom de l'environnement?" *Économie Rurale* 208/209 (1992): 137-41.

Blue Ribbon Study Commission on Agricultural Waste. *Report to the 1995 General Assembly of North Carolina, Regular Session.* Raleigh: May 1996.

Bressers, H.Th.A., and L.J. O'Toole Jr. "The Selection of Policy Instruments: A Network-Based Perspective." *Journal of Public Policy* 18 (1998): 213-39.

Browne, W.P. *Private Interests, Public Policy, and American Agriculture.* Lawrence: University Press of Kansas, 1988.

–. *Cultivating Congress: Constituents, Issues, and Interests in Agricultural Policymaking.* Lawrence: University Press of Kansas, 1995.

Busch, L., and A. Juska. "Beyond Political Economy: Actor Networks and the Globalization of Agriculture." *Review of International Political Economy* 4, 4 (1997): 688-708.

Cahart, P., L.-R. Bargard, A. Joly, C. Rogeau, J.-J. Benetiere, A. Gravaud, P. LeBail, and J.P.

Vogler. *Rapport d'évaluation sur la gestion et le bilan du programme de maîtrise des pollutions d'origine agricole*. Paris: Ministère de l'Économie, des Finances et de l'Industrie; Ministère de l'Aménagement du Territoire et de l'Environnement; Ministère de l'Agriculture et de la Pêche, 1999.

Callon M., P. Lascoumes, and Y. Barthe. *Agir dans un monde incertain*. Paris: Seuil, 2001.

Cameron, D., and D. Drache. "Outside the Macdonald Commission: Reply to Richard Simeon." *Studies in Political Economy* 26 (1988): 173-80.

Clayton, R., and J. Pontusson. "Welfare State Retrenchment Revisited: Entitlement Cuts, Public Sector Restructuring, and Inegalitarian Trends in Advanced Capitalist Societies." *World Politics* 51 (1998): 67-98.

Cochrane, W.W. *The Development of American Agriculture: A Historical Analysis*. Minneapolis: University of Minnesota Press, 1993.

Cohen, M., J. March, and J. Olsen. "A Garbage Can Model of Organizational Choice." *Administrative Science Quarterly* 17 (1972): 1-25.

Coleman, W.D. "Policy Communities and Policy Networks: Some Issues of Method." Paper prepared for presentation to the System of Government Conference, University of Pittsburgh, 1 November 1997.

–. "From Protected Development to Market Liberalism: Paradigm Change in Agriculture." *Journal of European Public Policy* 5, 4 (1998): 632-51.

–, M.M. Atkinson, and E. Montpetit. "Against the Odds: Retrenchment in Agriculture in France and the United States." *World Politics* 49, 4 (July 1997): 453-81.

–, and C. Chiasson. "State Power, Transformative Capacity and Adapting to Globalization: An Analysis of French Agricultural Policy 1960-2000." *Journal of European Public Policy* 9, 2 (2002): 168-85.

–, and A. Perl. "Internationalized Policy Environments and Policy Network Analysis." *Political Studies* 47, 4 (1999): 691-709.

–, and G. Skogstad, eds. *Policy Communities and Public Policy in Canada: A Structural Approach*. Mississauga: Copp Clark Pitman, 1990.

–, G. Skogstad, and M.M. Atkinson. "Paradigm Shifts and Policy Networks: Cumulative Change in Agriculture." *Journal of Public Policy* 16, 3 (1997): 273-301.

–, and S. Tangermann. "The 1992 CAP Reform, the Uruguay Round and the Commission: Conceptualizing Linked Policy Games." *Journal of Common Market Studies* 37, 3 (1999): 383-405.

Compston, H. "The End of National Policy Concertation? Western Europe Since the Single European Act." *Journal of European Public Policy* 5, 3 (1998): 507-26.

Coulomb, P. "Les conférences annuelles entre corporatisme et populisme." In *Les agriculteurs et la politique*, edited by P. Coulomb, H. Delorme, B. Hervieu, M. Jollivet, and P. Lacombe, 159-79. Paris: Presses de la Fondation nationale des sciences politiques, 1990.

Cummings, J. "Legislation Relating to Agricultural Waste and Regulations and Penalties on Pollution of Water Resources." Paper presented at the Agribusiness Law Conference, Campbell University, 1997.

Detlef, J. "Environmental Performance and Policy Regimes." *Policy Sciences* 31 (1998): 107-31.

Doern, G.B., and T. Conway. *The Greening of Canada: Federal Institutions and Decision*. Toronto: University of Toronto Press, 1994.

Dowding, K. "Model or Metaphor? A Critical Review of the Policy Network Approach." *Political Studies* 43 (1995): 136-58.

Dufour, J.-P. "Quand l'ozone des champs envahit les villes." *Le Monde*, 8 September 1998.

École nationale d'administration, under the direction of Lucien Chabasson. *L'aménagement de l'espace rural*. Volume 2. Paris: ENA, 1994.

Environment Canada. *A Guide to the New Canadian Environmental Protection Act*. Ottawa: Minister of Public Works and Government Services, 2000.

Environmental Protection Commission. "Adopted Rule Awaiting Publication: Chapter 65 Animal Feeding Operations." Des Moines: Government of Iowa, 1997.

European Commission. *CAP Working Notes: Agriculture and the Environment*. Special issue. Brussels: European Commission, 1997.

Evans, P.B. *Embedded Autonomy: States and Industrial Transformation.* Princeton: Princeton University Press, 1995.

Federal and Provincial Ministers of Agriculture. *National Environment Strategy for Agriculture and Agri-Food.* Ottawa: Agriculture and Agri-Food Canada, 1995.

Federal/Provincial/Territorial Advisory Committee on Canada's National Program of Action for the Protection of the Marine Environment from Land-based Activities. *Canada's National Program of Action for the Protection of the Marine Environment from Land-based Activities.* Ottawa: Minister of Public Works and Government Services Canada, 2000.

Finegold, K., and T. Skocpol. *State and Party in America's New Deal.* Madison: University of Wisconsin Press, 1995.

Food and Agriculture Organization of the United Nations. *The State of Food and Agriculture.* Rome: FAO, 1970.

–. *The State of Food and Agriculture.* Rome: FAO, 1990.

–. *Sustainability Issues in Agricultural and Rural Development Policies.* Volumes 1 and 2. Edited by F. Pétry. Rome: FAO, 1995.

Fouilleux, E. "Réforme de la PAC, accord du GATT: Quelles incidences sur les transferts financiers entre les Douze?" *Economie et Prévision* 117-18 (1995): 129-41.

–. *La cogestion à la française à l'épreuve de l'Europe: Le cas de la réforme de la Politique Agricole Commune.* Grenoble: Centre de recherche sur le politique, l'administration et le territoire, 1996.

Freeman III, A.M. "Economics, Incentives, and Environmental Regulation." In *Environmental Policy in the 1990's,* 3rd edition, edited by N.J. Vig and M.E. Kraft, 187-207. Washington, DC: Congress Quarterly Press, 1997.

Frieden, J.A., and R. Rogowski. "The Impact of the International Economy on National Policies: An Analytical Overview." In *Internationalization and Domestic Politics,* edited by R.O. Keohane and H.V. Milner, 25-47. Cambridge: Cambridge University Press, 1996.

Furuseth, O.J. "Restructuring of Hog Farming in North Carolina: Explosion and Implosion." *Professional Geographer* 49, 4 (1997): 391-403.

Goldstein, J. "The Impact of Ideas on Trade Policy: The Origins of U.S. Agricultural and Manufacturing Policies." *International Organization* 43, 1 (1989): 31-72.

Grant, W.P. "The Limits of the Common Agricultural Policy Reform and the Option of Renationalisation." *Journal of European Public Policy* 2 (1995): 1-18.

Green Cowles, M., J. Caporaso, and T. Risse, eds. *Transforming Europe: Europeanization and Domestic Change.* Ithaca: Cornell University Press, 2001.

Groupe de Bruges. *Agriculture: Un tournant nécessaire.* Paris: Éditions de l'aube, 1996.

Haas, P. "Do Regimes Matter? Epistemic Communities and Mediterranean Pollution Control." *International Organization* 43, 3 (1989): 377-403.

–. "Introduction: Epistemic Communities and International Policy Coordination." *International Organization* 46, 1 (1992): 1-36.

–, and E.B. Haas. "Learning to Learn: Improving International Governance." *Global Governance* 1 (1995): 260-62.

Hall, P. *Governing the Economy: The Politics of State Intervention in Britain and France.* New York: Oxford University Press, 1986.

Hansen, J.M. *Gaining Access: Congress and the Farm Lobby, 1919-1981.* Chicago: University of Chicago Press, 1991.

Harrison, K. *Federalism and Canadian Environmental Policy.* Vancouver: UBC Press, 1996.

Hatfield, J.L., and D.L. Karlen, eds. *Sustainable Agriculture Systems.* Boca Raton: Lewis Publisher, 1994.

Heidenheimer, A.J., H. Heclo, and C.T. Adams. *Comparative Public Policy: The Politics of Social Choice in America, Europe, and Japan.* 3rd edition. New York: St. Martin's Press, 1990.

Held, D., A. McGrew, D. Goldblatt, and J. Perraton. *Global Transformations: Politics, Economics and Culture.* Stanford: Stanford University Press, 1999.

Hirst, P. "The Global Economy: Myths and Realities." *International Affairs* 73 (1997): 409-25.

Hollingsworth, J.R., and R. Boyer, eds. *Contemporary Capitalism: The Embeddedness of Institutions.* Cambridge: Cambridge University Press, 1997.

Howlett, M. "Predictable and Unpredictable Windows: Institutional and Exogenous Correlates of Canadian Federal Agenda-Setting." *Canadian Journal of Political Science* 31, 3 (1998): 495-524.

–, and M. Ramesh. *Studying Public Policy: Policy Cycles and Policy Subsystems*. Toronto: Oxford University Press, 1995.

Inglehart, R. *Modernization and Postmodernization: Cultural, Economic, and Political Change in 43 Societies*. Princeton: Princeton University Press, 1997.

Institut français de l'environnement. *Agriculture et environnement: Les indicateurs*. Paris: IFEN, 1997.

Jenkins-Smith, H.C., and P.A. Sabatier. "The Dynamics of Policy-Oriented Learning." In *Policy Change and Learning: An Advocacy Coalition Approach*, edited by P.A. Sabatier and H.C. Jenkins-Smith, 41-56. Boulder: Westview Press, 1993.

Jung, C., K. Krutilla, and R. Boyd. "Incentives for Advanced Pollution Abatement Technology at the Industry Level: An Evaluation of Policy Alternatives." *Journal of Environmental Economics and Management* 30 (1996): 95-111.

Kaczmarek, B. "La politique communautaire de l'eau." *Aménagement et Nature* 124 (March 1997): 25-36.

Keeler, J.T.S. *The Politics of Neocorporatism in France: Farmers, the State, and Agricultural Policy-Making in the Fifth Republic*. Oxford: Oxford University Press, 1987.

Killpatrick, K. "Concern Grows about Pollution from Megafarms." *Globe and Mail*, 30 May 2000.

Kincaid, J. "Intergovernmental Costs and Coordination in U.S. Environmental Protection." In *Federalism and the Environment: Environmental Policymaking in Austria, Canada and the United States*, edited by K.M. Holland and F.L. Morton, 79-101. Westport: Greenwood Press, 1996.

Kingdon, J.W. *Agendas, Alternatives, and Public Policies*. New York: HarperCollins College Publishers, 1995 [1984].

Kohler-Koch, B. "Catching up with Change: The Transformation of Governance in the European Union." *Journal of European Public Policy* 3, 3 (1996): 359-80.

Lajoie, N., and F. Blais. "Une réconciliation est-elle possible entre l'environnement et le marché? Une évaluation critique de deux tentatives." *Politique et Sociétés* 18, 3 (1999): 49-77.

Lascoumes, P. *L'éco-pouvoir: Environnement et politiques*. Paris: Édition la découverte, 1994.

Le Doux, J. "Les bonnes recettes de deux producteurs bretons." *Ouest France*, 13 May 1997.

Ledru, M., for the Conseil économique et social. "L'espace rural entre protection et contraintes." *Journal Officiel de la République Française*. Paris: Avis et rapports du Conseil économique et social, 1994.

Lemaire, D. "The Stick: Regulation as a Tool of Government." In *Carrots, Sticks and Sermons: Policy Instruments and Their Evaluation*, edited by M.-L. Bemelmans-Videc, R.C. Rist, and E. Vedung, 66-67. New Brunswick: Transaction Publishers, 1998.

Lemieux, V. *L'étude des politiques publiques: Les acteurs et leur pouvoir*. Sainte-Foy: Les Presses de l'Université Laval, 1995.

"Les mesures agri-environnement." *Entraid'Ouest* supplement (June 1997): 5.

Lotspeich, R. "Comparative Environmental Policy: Market-Type Instruments in Industrialized Capitalist Countries." *Policy Studies Journal* 26, 1 (1998): 85-104.

MacAvoy, G.E. *Controlling Technocracy: Citizen Rationality and the Nimby Syndrome*. Washington, DC: Georgetown University Press, 1999.

Majone, G. *Evidence, Argument, and Persuasion in the Policy Process*. New Haven: Yale University Press, 1989.

Mancuso, M., M.M. Atkinson, A. Blais, I. Greene, and N. Nevitte. *A Question of Ethics: Canadians Speak Out*. Toronto: Oxford University Press, 1998.

Marks, G., L. Hooghe, and K. Blank. "European Integration from the 1980s: State-Centric v. Multi-Level Governance." *Journal of Common Market Studies* 34, 3 (1996): 341-78.

Marsh, D., and M. Smith. "Understanding Policy Networks: Towards a Dialectical Approach." *Political Studies* 48 (2000): 4-21.

Ministère de l'Environnement et de la Faune. *Le réseau-rivières: Un baromètre de la qualité de nos cours d'eau*. Quebec: MEF, 1991.
Ministry of Agriculture and Food. "Nutrient Management Act." News release, 28 June 2002.
Montpetit, E. "Corporatisme québécois et performance des gouvernants: Analyse comparative des politiques environnementales en agriculture." *Politique et Sociétés* 18, 3 (1999): 79-98.
–. "Europeanization and Domestic Politics: Europe and the Development of a French Environmental Policy for the Agricultural Sector." *Journal of European Public Policy* 7, 4 (2000): 576-92.
–. "Sound Science and Moral Suasion, Not Regulation: Facing Difficult Decisions on Agricultural Non-Point Source Pollution." In *Canadian Environmental Policy: Context and Cases,* 2nd edition, edited by D. Van Nijnatten and R. Boardman, 274-85. Don Mills: Oxford University Press, 2002.
–, and W.D. Coleman. "Policy Communities and Policy Divergence in Canada: Agro-Environmental Policy Development in Quebec and Ontario." *Canadian Journal of Political Science* 32, 4 (1999): 691-714.
Morgan, R.J. *Governing Soil Conservation: Thirty Years of the New Decentralization*. Baltimore: Johns Hopkins Press, 1965.
Muller, P. *Le technocrate et le paysan*. Paris: Éditions ouvrières, 1984.
–. "Gouvernance européenne et globalisation." *Revue Internationale de Politique Comparée* 6 (1999): 701-17.
National Environmental Dialogue on Pork Production. *Comprehensive Environmental Framework for Pork Production Operations*. Washington, DC: America's Clean Water Foundation, 1997.
Nolet, J. *Étude comparative de différentes réglementations concernant les nuisances en agriculture*. Quebec: Rapport remis au ministère de l'Environnement et de la Faune, 1996.
North, D.C. *Institutions, Institutional Change and Economic Performance*. Cambridge: Cambridge University Press, 1990.
O'Connor, D.R. *Report of the Walkerton Inquiry: The Events of May 2000 and Related Issues*. Toronto: Publication Ontario, 2002.
–. *Report of the Walkerton Inquiry: A Strategy for Safe Drinking Water*. Toronto: Publication Ontario, 2002.
Ontario Farm Environmental Coalition. *Our Farm Environmental Agenda*. Ontario: OFEC, 1992.
Organization for Economic Cooperation and Development. *Economic Accounts for Agriculture*. Paris: OECD, published annually.
–. *Environmental Performance in OECD Countries: Progress in the 1990s*. Paris: OECD, 1996.
–. *Environmental Performance Review: France*. Paris: OECD, 1997.
–. *OECD Environmental Data: Compendium 1997*. Paris: OECD, 1997.
Paarlberg, R. "Agricultural Policy Reform and the Uruguay Round: Synergistic Linkage in a Two-Level Game?" *International Organization* 51, 3 (1997): 413-44.
Parikh, J.K. "Introduction." In *Sustainable Development in Agriculture,* edited by J.K. Parikh, 1-17. Dordrecht: Maritimus Nijhoff Publishers, 1988.
Peterson, J. "Policy Networks and European Union Policy Making: A Reply to Kassim." *West European Politics* 18, 1 (1995): 389-407.
–. "States, Societies and the European Union." *West European Politics* 20, 4 (1997): 1-23.
Pierre, J., and B.G. Peters. *Governance, Politics and the State*. New York: St. Martin's Press, 2000.
Pierson, P. "When Effect Becomes Cause: Policy Feedback and Political Change." *World Politics* 45 (1993): 595-628.
–. *Dismantling the Welfare State? Reagan, Thatcher, and the Politics of Retrenchment*. Cambridge: Cambridge University Press, 1994.
–. "The New Politics of the Welfare State." *World Politics* 48 (1996): 143-79.
–. "Increasing Returns, Path Dependence, and the Study of Politics." *American Political Science Review* 94, 2 (2000): 251-67.

Polanyi, Karl. *Great Transformation.* Boston: Beacon Press, 1957 [1944].

Powell, D. "Deadly Bacteria Show up in Surprising Places." *Globe and Mail,* 25 January 1997.

Putnam, R.D. "Diplomacy and Domestic Politics." *International Organization* 42 (1988): 427-61.

–. *Making Democracy Work: Civic Traditions in Modern Italy.* Princeton: Princeton University Press, 1993.

–, S.J. Pharr, and Russell J. Dalton. "Introduction: What's Troubling the Trilateral Democracies?" In *Disaffected Democracies: What's Troubling the Trilateral Countries?* edited by S.J. Pharr and R.D. Putnam, 3-27. Princeton: Princeton University Press, 2000.

Rabe, B.G. "Power to the States: The Promise and Pitfalls of Decentralization." In *Environmental Policy in the 1990s,* 3rd edition, edited by N.J. Vig and M.E. Kraft, 31-52. Washington, DC: Congressional Quarterly Press, 1997.

–. "Federalism and Entrepreneurship: Explaining American and Canadian Innovation in Pollution Prevention and Regulatory Integration." *Policy Studies Journal* 27 (1999): 288-306.

–, and J.B. Zimmerman. "Beyond Environmental Regulatory Fragmentation." *Governance: An International Journal of Policy and Administration* 8, 1 (1995): 58-77.

Rainelli, P. "Intensive Livestock Production in France and Its Effects on Water Quality in Brittany." In *Toward Sustainable Agricultural Development,* edited by M.D. Young, 126-44. London: Belhaven Press, 1992.

–, and D. Vermersch. "Thematic Network on the CAP and the Environment in the European union: A French Report." Unpublished paper, 1997.

Rapport de la France à la Commission du développement durable des Nations unies. Paris: Ministère de l'Aménagement du Territoire et de l'Environnement et Ministère des Affaires étrangères, 2000.

Rhodes, R.A.W. *Understanding Governance: Policy Networks, Governance, Reflexivity and Accountability.* Buckingham: Open University Press, 1997.

Ribaudo, M.O. "Lessons Learned about the Performance of USDA Agricultural Nonpoint Source Pollution Programs." *Journal of Soil and Water Conservation* 53, 1 (1998): 4-10.

Ricard, F. *La génération lyrique.* Montreal: Boréal, 1992.

Risse, T. "Let's Argue! Communicative Action in World Politics." *International Organization* 54, 1 (2000): 1-39.

Risse-Kapen, T. "Exploring the Nature of the Beast: International Relations Theory and Comparative Policy Analysis Meet the European Union." *Journal of Common Market Studies* 34, 1 (1996): 53-80.

Rousso, A. "Le droit du paysage: Un nouveau droit pour une nouvelle politique." *Courrier de l'Environnement de l'INRA* 26 (1995), <http://www.inra.fr/Internet/Produits/dpenv/roussc26.htm> (February 2003).

Ruggie, J. "International Regimes, Transactions, and Change: Embedded Liberalism in the Postwar Economic Order." *International Organization* 36 (1982): 379-415.

Salmon, J. "Environnement et élevage." *L'Information Agricole* 683 (November 1995): 17.

Scharpf, F.W. *Crisis and Choice in European Social Democracy.* Ithaca: Cornell University Press, 1987.

–. *Games Real Actors Play: Actor-Centered Institutionalism in Policy Research.* Boulder: Westview Press, 1997.

–. *Governing in Europe: Effective and Democratic?* Oxford: Oxford University Press, 1999.

Schneider, A.L., and H. Ingram. *Policy Design for Democracy.* Lawrence: University Press of Kansas, 1997.

Secrétariat du Conseil du trésor. *Gérer la réglementation au Canada.* Ottawa: Ministre des Approvisionnements et Services, 1996.

Simeon, R. "Inside the Macdonald Commission." *Studies in Political Economy* 22 (1987): 167-79.

Skogstad, G. *The Politics of Agricultural Policy-Making in Canada.* Toronto: University of Toronto Press, 1987.

–. "The Farm Policy Community and Public Policy in Ontario and Quebec." In *Policy*

Communities and Public Policy in Canada, edited by William Coleman and Grace Skogstad, 59-92. Toronto: Copp Clark Pitman, 1990.
–. "Agricultural Policy." In *Border Crossings: The Internationalization of Canadian Public Policy,* edited by G.B. Doern, L.A. Pal, and B.W. Tomlin, 143-64. Toronto: Oxford University Press, 1996.
–. "Ideas, Paradigms and Institutions: Agricultural Exceptionalism in the European Union and the United States." *Governance: An International Journal of Policy and Administration* 11, 4 (1998): 463-90.
Spedding, C.R.W. *An Introduction to Agricultural Systems.* London: Elsevier Applied Science, 1988.
Sproule-Jones, M. *Restoration of the Great Lakes: Promises, Practices, Performances.* Vancouver: UBC Press, 2002.
Standing Committee on Agriculture, Fisheries and Forestry. *Soil at Risk.* Ottawa: Senate of Canada, 1984.
Stein, J.G. *The Cult of Efficiency.* Toronto: Anansi, 2001.
Steinmo, S. *Taxation and Democracy: Swedish, British, and American Approaches to Financing the Modern State.* New Haven: Yale University Press, 1993.
Stith, P. "The Smell of Money." *News and Observer,* 24 February 1995.
–, J. Warrick, and M. Sill. "Boss Hog: North Carolina's Pork Revolution." *News and Observer,* 12 May 1996, reprinted edition.
Strange, S. *The Retreat of the State: The Diffusion of Power in the World Economy.* Cambridge: Cambridge University Press, 1996.
Streeck, W. "From National Corporatism to Transnational Pluralism: European Interest Politics and the Single Market." In *Participation in Public Policy-Making: The Role of Trade Unions and Employers' Associations,* edited by T. Treu, 97-126. Berlin: Walter de Gruyter, 1992.
–, and P.C. Schmitter. "Community, Market, State – and Associations? The Prospective Contribution of Interest Governance to Social Order." In *Private Interest Government,* edited by W. Streeck and P.C. Schmitter, 1-29. London: Sage Publications, 1985.
Sutton, A.L., and D.D. Jones. "Animal Waste Management Regulations: A Look into the Future." Unpublished manuscript, 1997.
Swine Odor Task Force. *Options for Managing Odor.* Raleigh: North Carolina Research Service, North Carolina State University, 1995.
Thelen, K., and S. Steinmo. "Historical Institutionalism in Comparative Politics." In *Structuring Politics: Historical Institutionalism in Comparative Analysis,* edited by S. Steinmo, K. Thelen, and F. Longstreth, 1-32. New York: Cambridge University Press, 1992.
Thurman, W.N. *Assessing the Environmental Impact of Farm Policies.* Washington, DC: AEI Press, 1995.
Tsebelis, G. "Decision Making in Political Systems: Veto Players in Presidentialism, Parliamentarism, Multicameralism and Multipartism." *British Journal of Political Science* 25 (1995): 289-325.
United States Department of Agriculture. *Agricultural Statistics 1961.* Washington, DC: United States Printing Office, 1961.
–. *Agricultural Statistics 1997.* Washington, DC: United States Printing Office, 1997.
United States Environmental Protection Agency. *Guide Manual on NPDES Regulations for Concentrated Animal Feeding Operations.* Washington, DC: EPA, December 1995.
–. *Clean Water Action Plan: Restoring and Protecting America's Waters.* Washington, DC: EPA, February 1998.
–. "EPA to Better Protect Public Health and the Environment from Animal Feeding Operations." Headquarters press release, 5 March 1998.
–. *Strategy for Addressing Environmental and Public Health Impacts from Animal Feeding Operations.* Washington, DC: EPA, 1999.
–. *Proposed Revisions to CAFO Regulations.* Washington, DC: EPA, 2001.
"Un tiers des ressources en eau potable sous la menace des nitrates." *Le Monde,* 10 June 1997.
Unland, K. "Pig Odour Raises a Stink in Quebec." *Globe and Mail,* 21 December 1996.
Valo, M. "L'État mis en cause pour son laxisme face à la pollution des eaux par les nitrates." *Le Monde,* 19 April 2001.

van der Ploeg, J.D., and G. van Dijk, eds. *Beyond Modernization: The Impact of Endogenous Rural Development*. Assen: Van Gorcum, 1995.

van Waarden, F.V. "Dimensions and Types of Policy Networks." *European Journal of Political Research* 21 (1992): 29-52.

Vogel, D. *National Styles of Regulation: Environmental Policy in Great Britain and the United States*. Ithaca: Cornell University Press, 1986.

–. "Representing Diffuse Interests in Environmental Policymaking." In *Do Institutions Matter? Government Capabilities in the United States and Abroad,* edited by R.K. Weaver and B.A. Rockman, 237-71. Washington, DC: The Brookings Institution, 1993.

Weersink A., J. Livernois, J.F. Shogren, and J. Shortle. "Economics Instruments and Environmental Policy in Agriculture." *Canadian Public Policy* 24, 3 (1998): 309-28.

Weiss, L. *The Myth of the Powerless State*. Ithaca: Cornell University Press, 1998.

Williamson, O.E. *Markets and Hierarchies: Analysis and Antitrust Implications*. New York: The Free Press, 1975.

Winfield, M. "The Ultimate Horizontal Issue: The Environmental Policy Experiences of Alberta and Ontario, 1971-1993." *Canadian Journal of Political Science* 27 (1994): 129-52.

Wolf Jr., C. "Carences du marché et carences hors-marché: Comparaison et évaluation." *Politiques et Management Public* 5, 1 (1987): 57-84.

–. *Market or Governments: Choosing between Imperfect Alternatives*. Cambridge: MIT Press, 1988.

Index

Note: "(t)" after a number indicates a table.

Actor constellations: balance of cohesion and diversity, 41-2, 46, 49-50, 109, 113, 122; compared with epistemic communities, 43; composition, 41-3; favourable for policy-making performance, 41-6, 105, 109; impact of intergovernmentalism, 49-51, 119; impact of multilevel governance, 50(t), 52, 64, 119-20; intergovernmentalism and patterns of exclusion, 50, 119; in new politics of welfare state, 48, 116-17; power of joint action, 41-2, 109, 110, 122; side payments, 42-3, 46, 68-9; truth-oriented dialogue, 42, 111; veto players, 43. *See also* Civil society actors; State actors

Actor constellations, Canada: Agriculture Ministry responsible for environment, 98; cause of poor policy-making performance, 90, 96-7, 112, 118, 120; Christian Farmers Federation of Ontario (CFFO), 100-1; cohesiveness, 99, 103-4, 112, 121; compared with France and US, 112-13; farmers' dissatisfaction with federal government, 98; federal agencies weak, 97-8, 112, 121; federal "regulation for regulations," 97-8, 112; federal reliance on education, 90, 96, 98-9, 112; impact of Farm Registration and Farm Organizations Funding Act, 101-2, 136n37; National Farmers Union (NFU), 100-1; Ontario Farm Environmental Coalition (OFEC), 101-2; Ontario Federation of Agriculture, 100-1; Ontario Ministry of Agriculture, power of, 102, 103, 118; Ontario Ministry of Environment, power of, 102, 103, 118; Ontario's clientelist policy network, 102, 112-13, 118; Ontario's cohensive group, 99-103, 136n28; "Our Farm Environmental Agenda," 101; provincial cohesive agriculturalist group, 100, 112

Actor constellations, France: changes in France (1980s to 1990s), 64-7, 69-70, 109-10; *cogestion,* 67; compared with US and Canada, 83, 112-13; Confédération paysanne, 66-7; CORPEN (1980s to 1990s), 57, 59, 64-7; and corporatist network structure, 17, 55, 67-8, 69-70, 109-11, 113, 120; FNSEA, 54-5, 66, 67-8, 70, 84, 110; ministries of environment and agriculture, 64-6, 67, 68-70, 109-10; side payments, 55, 68-9, 111

Actor constellations, United States: compared with France and Canada, 83, 111, 112-13; EPA *(see* Environmental Protection Agency); federal advocacy coalitions, 84-6; in federal state-directed policy networks, 85-6, 112; fragmentation of interest groups, 83, 88; state players, 87-8, 111; in state pluralist policy networks, 87-8, 111, 113; support for farm bills, 86; US Department of Agriculture (USDA), 72, 86

Agenda setting: agricultural pollution on political agenda, 13-15, 38, 106, 115-16; definition, 38; distrust of policy makers' ability, 38-40, 114-15; garbage can model, 38; theories of agenda setting, 38-40, 114-16; windows of opportunity, 38-40, 114-15

Agri-Environmental Indicator Project (Canada), 92(t), 93

Agri-environmental Measures of CAP, 62
Agricultural Code of Practice (Ontario), 94
Agricultural economists, 27-8
Agricultural Environmental Stewardship Initiative (AESI, Canada), 92(t), 93
Agricultural pollution: cross-media pollution, 14, 25-6, 31-2, 33; knowledge re solutions, 9; seriousness of problem, 13-15, 29-30, 38
Agricultural scientists, 27-8
Agricultural sector: agricultural exceptionalism, 26-7; similarity of Canada, France, and US, 16; state involvement, 9-10; strong organizations, 9-10
Agriculture Canada, 14, 98, 112
Agriculture system analysts, 24(t), 27-8, 29, 31-2, 34(t)
Agro-environmental policy development: agricultural exceptionalism since WWII, 26-7; agriculture system analysts, 24(t), 27-8, 29, 31-2, 34(t); alternative agriculture, 24(t), 28, 29, 32, 34(t); Canada, France, US compared, 71, 72, 83, 90, 94, 96, 103-4, 106-9, 113, 120-1; characteristics of policies used for evaluation, 30-3, 34(t), 35; Clean Water Action Plan (US), 86-7; ecological scientists, 24(t), 26-7, 29, 31-2, 34(t); economic viability of agriculture, 32-3, 34(t), 35; epistemic communities on seriousness of pollution, 29-30, 33, 34(t); in France in 1980s, 56, 57-9, 63-4; in France in 1990s, 59-64; new politics period in US and OECD countries, 48-9; principles for evaluating policy success, 12; US disconnect between federal and state networks, 33, 71, 88-9, 108, 111-12; US farm bills and agricultural policy, 76, 86; US federal policies, 72, 73-9, 82-3; US state policies, 72, 73, 74(t), 77(t), 79-83, 87-8; utilitarian view of success, 10-11, 121. *See also* Policy instruments; *entries beginning with* Policy-making performance
Air pollution: cross-media pollution, 14, 25-6, 31-2, 33
Alternative agriculture, 24(t), 28, 29, 32, 34(t)
Article 19 (European Community policy), 57-8
Association nationale pour le développement agricole (ANDA), 59
Atkinson, Michael M., 8

Best Management Practices project (Ontario), 92(t), 95
Blair House Agreement (1993), 54, 66, 130n5, 132n52
Bressers, Hans Th.A., 7-8
Bureaucracies: distrust of, 4, 5, 37, 118; need for strong bureaucracies, 46, 70, 110, 118; and policy-making performance, 5

CAFOs (concentrated animal feeding operations): EPA definition, 73; "Pork Dialogue," 85; *Proposed Revisions to CAFO Regulations* (EPA, 2001), 73, 75; US federal regulations, 73, 75; US state regulations, 79-80, 81(t), 82
Canada: agricultural sector, similarity to France and US, 16; choice for comparative study, 15-18; governance and policy network, 16-17; recognition of agricultural pollution problem, 13-15, 38, 106, 115; water and air pollution, 14-15. *See also* Actor constellations, Canada; Policy instruments, Canada
Canadian Adaptation and Rural Development (CARD) fund, 92(t), 93
Canadian Environmental Protection Act (CEPA), 91, 92(t), 93-4
Centre d'étude pour un développement agricole plus autonome (CEDAPA), 28
Civil society actors: in clientelist networks, 46, 112-13, 118; incorporatist networks, 44, 67, 109-10, 117-18; excluded in intergovernmentalism, 50, 119; for governed interdependence, 45, 50, 109; influence in coalitions, 101; in new politics of welfare state, 48, 116-17; role in governance, 4. *See also entries beginning with* Actor constellations
Clean Up Rural Beaches (CURB) program (Ontario), 102
Clean Water Act (US, 1972), 73, 74(t)
Clean Water Action Plan (US, 1998), 73, 74(t), 86-7
Clean Water Foundation (US), 85
Clientelist networks: balance of power of players, 44-6, 112, 118; Ontario's networks, 90, 100-3, 104, 112-13, 118; structure's influence on policy-making performance, 45-6, 118
Coastal Zone Act (US), 74(t), 76, 78
Code rural (France), 57(t), 58
Coleman, William D., 8, 41, 44, 49, 51
Common Agricultural Policy (CAP, Europe), 14, 54, 62, 66

Concentrated animal feeding operations. *See* CAFOs
Confédération paysanne (France), 66-7
Conservation Authorities Act (Ontario), 92(t), 95
Conservation Compliance (US), 74(t), 78
Conservation Reserve Enhancement Program (CREP, US), 76, 78, 83, 107
Conservation Reserve Program (CRP, US), 74(t), 76, 78, 83, 107
Contrat territorial d'exploitation (CTE, France), 64, 111, 117
Conway, Thomas, 104
CORPEN (Comité d'orientation pour la réduction de la pollution des eaux par les nitrates), 57, 59, 60(t), 64-6, 107
Corporatism, 52
Corporatist networks: actor constellations leading to strong policy-making performance, 44, 45, 46, 67, 105, 109-10, 117-18; adaptation to regional or international pressure, 8; balance of power of actors, 44-5, 46, 67, 109, 117-18; corporatist tradition in France, 17, 55, 67-8, 69-70, 109-11, 113, 120; impact of multilevel governance, 50, 52-3, 119; most conducive to governed interdependence, 8, 45, 46, 67, 105, 109
Corruption: as cause of distrust, 3-4
Crisis of confidence. *See* Distrust
Cryptosporidium (parasite), 14

Dalton, Russell J., 3-4, 10, 15
DEXEL *(diagnostic environnemental d'exploitation),* 61
Discretionary windows of opportunity, 39, 114
Distrust: agenda-setting ability of policy makers, 38-40, 114-15; of bureaucracies, 4, 5, 37, 118; concerns about governance, 4; consequences of distrust, 5; existence across different countries, 15; impact of governance theories, 4-5, 37, 105; intergovernmentalism's exclusion of civil society actors, 50, 119; internationalism and governance inadequacy, 20, 37, 119; motives of policy makers, 3-5, 37, 114; new politics of the welfare state, 47-9, 116-18; reasons for, 3-4
Doern, G. Bruce, 104

Ecological scientists, 24(t), 26-7, 29, 31-2, 34(t)
Environmental Farm Plan (EFP) program (Ontario), 92(t), 94
Environmental Protection Act (Ontario), 92(t), 94, 99
Environmental Protection Agency (EPA, US): and Clean Water Action Plan, 86-7; Coastal Zone Act (Section 319 Program), 74(t), 76, 78; cross-media pollution, 14; *Proposed Revisions to CAFO Regulations* (2001), 73, 75; regulations re CAFOs, 73, 75; weak player in US actor constellation, 72, 86-7
Environmental Quality Incentives Program (EQIP, US), 74(t), 76, 83, 107
Epistemic communities: absence of consensus on preferred policy instruments, 29; agreement on seriousness of agricultural pollution, 29-30, 35, 38; agricultural economists, 27-8; agricultural scientists, 27-8; agriculture system analysts, 24(t), 27-8, 29, 31-2, 34(t); alternative agriculture, 24(t), 28, 29, 32, 34(t); approval for various policy instrument types, 24(t); characteristics of agro-environmental policy, 31-3, 34(t), 35; compared with actor constellations, 43; compared with policy networks, 25; definition, 25; ecological scientists, 24(t), 26-7, 29, 31-2, 34(t); importance of geographic proximity, 26; need for effective policy instruments, 31-3, 34(t), 35; principles for evaluating agro-environmental policy success, 12; science as basis for policy, 29, 127n39
European Agricultural Guarantee and Guidance Fund (EAGGF), 62
European Commission, 64, 65-6
European Community: Article 19 as incentive policy instrument, 57-8; European Agricultural Guarantee and Guidance Fund (EAGGF), 62; extensification as policy instrument, 57(t), 58, 131n15; Nitrate Directive adopted, 59, 66; recognition of agricultural pollution problem, 106
European Union: Common Agricultural Policy (CAP), 14; impact on legitimacy of French environment ministry, 55; multilevel governance policy environment, 49, 51-2, 119

FAOSTAT (Food and Agriculture Organization database), 16
Farm Practices Protection Act (Ontario), 92(t), 94-5
Farm Registration and Farm

Organizations Funding Act (Ontario), 101-2, 136n37
Farmers. *See* Agricultural sector
Fédération nationale des syndicats d'exploitants agricoles (FNSEA): stance in water agency discussions, 68-9; veto player in France, 54-5, 68, 84, 110
Ferti-Mieux, 59
Fertilizers Act (Canada), 91, 92(t)
Fouilleux, Eve, 67
France: agricultural sector compared with Canada and US, 16; choice for comparative study, 15-18; corporatist policy network, 17, 55, 67-8, 69-70, 109-11, 113, 120; and European Common Agricultural Policy, 14; governance and policy network, 16-17, 119-20; Hénin Report (1980), 14, 106; move to high agro-environmental performance, 54-5, 63-4; and new politics of the welfare state, 117; recognition of pollution problem, 13-15, 38, 54, 106, 115; water and air pollution, 14-15. *See also* Actor constellations, France; Policy instruments, France
Free trade, 49, 50

Globalization. *See* Internationalization
Governance: criticisms of public, 4, 20; description, 4; France, Canada, US compared, 16-17; theories that inspire distrust, 4-5, 37. *See also* Multilevel governance
Great Lakes basin: cross-media pollution, 25-6

Haas, Peter M., 25, 26, 28
Hénin Report (France, 1980), 14, 106
Hirst, Paul, 121
Hog production: advocacy coalition for "Pork Dialogue," 84-5; odour problem, 14, 125n51; PMPOA program of France, 60(t), 61, 132n41; US Congressional bills re operations, 75; US state livestock regulations, 79-80, 81(t), 82
Howlett, Michael, 39-40, 114

Inglehart, Ronald, 3, 9
Ingram, Helen, 4, 43
Institut français de l'environnement, 14
Institutional (routine) windows of opportunity, 39, 114
Interest groups, 4, 5, 20, 37. *See also entries beginning with* Actor constellations
Intergovernmentalism: exclusion of civil society actors, 50, 119; impact on policy networks, 49-52, 119; internationalized policy environment, 49, 50
Internationalization, 8, 20, 37, 49, 119-22
Internationalized policy environments: impact on Canada's policy-making performance, 104; intergovernmental negotiations *(see* Intergovermentalism); loose couplings, 49, 50(t); multilevel governance *(see* Multilevel governance); policy networks, 49-52; self-regulatory and private regimes, 49, 50(t)
Iowa: livestock regulations, 79-80, 81(t), 82
Issue networks, 44, 45

Jenkins-Smith, Hank C., 84
Jones, Don, 79

Kingdon, John W., 38, 114
Kohler-Koch, Beate, 52

Land Stewardship I and II (Ontario), 100
Land use planning: in France in 1980s, 57(t), 58
Landscape Policy (France), 60(t), 61
Law for the Protection of Nature (France), 57(t), 58
Law on Classified Infrastructures (France), 57(t), 58, 60
Legitimacy, input- and output-oriented, 6-7, 105, 121
Links between Actions for the Development of the Rural Economy (LEADER), 63, 132n37
Livestock: contribution to global warming, 14; manure management, 14; odour problem, 14, 125n51; US federal policy instruments, 73, 75; US state regulations, 79-80, 81(t), 82, 87. *See also* CAFOs (concentrated animal feeding operations); Hog production
Loose couplings, 49, 50(t)
Lotspeich, Richard, 35

McAvoy, Gregory E., 7
Majone, Giandomenico, 25, 26
Marks, Gary, 64
"Massachusetts Miracle Fallacy," 11-12
Milwaukee, Wisconsin,: contamination in drinking water, 14
Multilevel governance: in European Union, 49, 51-2, 119-20; impact on

actor constellations, 50(t), 52, 64, 119-20; impact on corporatist policy networks, 50, 52, 119; internationalized policy environment, 49

National Farmers Union (NFU, Ontario), 100-1
National Pollutant Discharge Elimination System (NPDES, US) permits, 73, 81(t)
National Pork Producers Council (NPPC, US), 84-5
National Program of Action for the Protection of the Marine Environment from Land-Based Activities (NPA, Canada), 91, 92(t), 93
Natural Resources Conservation Service (NRCS, US), 86
New politics of welfare state, 47-9, 116-18
News media: role in crisis of confidence, 3
Nitrate Directive, 59, 60(t), 66, 68, 107
North, Douglass C., 47
North American Free Trade Agreement (NAFTA), 49, 50
North Carolina: livestock regulations, 79-80, 81(t), 82, 87
Nutrient Management Act (Ontario, 2002), 102-3

O'Connor, Dennis, 96, 102-3, 113
Oklahoma: livestock regulations, 79-80, 81(t), 82
Ontario Farm Environmental Coalition, 101-2
Ontario Federation of Agriculture (OFA), 100-1
Ontario Planning Act, 92(t), 95
Ontario Water Resources Act, 92(t), 95
Organization for Economic Cooperation and Development (OECD), 15, 22, 48-9
O'Toole, Laurence J., Jr., 7-8

Pal, Leslie A., 22
Perl, Anthony, 49, 51
Peters, B. Guy, 4
Pfiesteria (microbe), 15
Pharr, Susan J., 3-4, 10, 15
Pierre, Jon, 4
Pierson, Paul, 47-9, 116-18
Plans de développement durable (PDD, France), 60(t), 61-2
Policy instruments: absence of consensus by epistemic communities, 23, 24(t), 29, 33, 34(t); Canada, France, US compared, 106-9; command-and-control regulations, 24(t); comprehensiveness, 31-2, 33, 34(t), 35, 106, 107-8; cross-compliance, 24(t); definition, 55; economic viability of agriculture, 32-3, 34(t), 35, 106, 108; education/moral suasion, 24(t); endogenous, 24(t); financial incentives/subsidies, 24(t); intrusiveness, 31, 33, 34(t), 35, 106-7; network characteristics and choice of policy instruments, 7-8; reformative, 24(t); self-regulation/self-governance, 24(t); types, 23, 24(t). *See also entries below beginning with* Policy instruments
Policy instruments, Canada (federal): Agri-Environmental Indicator Project, 92(t), 93; Agricultural Environmental Stewardship Initiative (AESI), 92(t), 93; Canadian Adaptation and Rural Development (CARD) fund, 92(t), 93; Canadian Environmental Protection Act (CEPA), 91, 92(t), 93-4; compared with France and US, 106-9; comprehensiveness, 96(t), 107-8; cross-compliance, 96(t); and economic viability of agriculture, 108; educational, 91, 92(t), 93-4, 96, 98, 107, 112; Fertilizers Act, 91, 92(t); financial incentives, 92(t), 93, 96, 106-7; income support for farmers lacking, 98; internationalization, impact of, 104; intrusiveness, 96(t), 106-7; National Program of Action for the Protection of the Marine Environment from Land-Based Activities (NPA), 91, 92(t), 93; regulatory, 91, 92(t), 93-4, 96, 106-7, 112; St. Lawrence Action Plan Vision 2000, 91, 92(t), 93, 135n8
Policy instruments, Canada (provincial): Agricultural Code of Practice (Ontario), 94; Best Management Practices project (Ontario), 92(t), 95; Clean Up Rural Beaches (CURB) program, 102; comprehensiveness, 96(t); Conservation Authorities Act (Ontario), 92(t), 95; cross-compliance, 92(t), 95, 96(t); educational, 92(t), 95, 96(t); Environmental Farm Plan (EFP) program (Ontario), 92(t), 94; Environmental Protection Act (Ontario), 92(t), 94, 99; Farm Practices Protection Act (Ontario), 92(t), 94-5; financial incentives, 92(t), 95, 96(t); intrusiveness, 96(t); Land Stewardship I and II, 100; Nutrient Management Act (Ontario, 2002), 102-3; Ontario Planning Act, 92(t), 95; Ontario Water

Resources Act, 92(t), 95; Quebec's use of regulations and incentives, 94; regulatory, 92(t), 94, 96(t), 102-3; self-regulatory, 96(t), 99-100; Soil and Water Environmental Enhancement Program (SWEEP), 99; voluntary measures predominant, 92(t), 94-6

Policy instruments, France: Agri-environmental Measures of CAP, 62; command-and-control regulations, 56, 59-60, 63, 106, 107; compared with US and Canada, 72, 83, 106-9; comprehensiveness, 56, 59, 63-4, 107; Contrat territorial d'exploitation (CTE), 64, 111, 117; CORPEN, 57, 59, 60(t), 64-6, 107; cross-compliance, 56, 60(t), 62; definition, 55; economic protection for farmers, 63, 108-9; education/moral suasion, 56, 59-60, 63; endogenous, 56, 60(t), 62-3; financial incentives/subsidies, 56, 60(t), 61, 63, 106; intrusiveness, 56, 59, 63-4, 106; Landscape Policy, 60(t), 61; Law for the Protection of Nature, 57(t), 58; Law on Classified Infrastructures, 57(t), 58, 60; move to high agro-environmental performance, 54-5, 63-4, 69-70; Nitrate Directive, 59, 60(t), 66, 68, 107; Plans de développement durable (PDD), 60(t), 61-2; Programme de maîtrise des pollutions d'origine agricole (PMPOA), 60(t), 61, 63, 68-9, 108, 132n41; reformative, 56, 60(t), 61, 63; rural development and quality policies, 60(t), 62-3; in 1980s, 55, 56, 57-9, 63-4, 69; in 1990s, 55, 59-64, 69; side payments, 55, 68-9, 111; water policy (1992), 59-60; Zones d'excédent structurel (ZES), 60

Policy instruments, United States (federal): CAFO (concentrated animal feeding operations) policies, 73, 74(t), 75; Clean Water Act (1972), 73, 74(t); Clean Water Action Plan (1998), 73, 74(t), 86-7; Coastal Zone Act (Section 319 Program), 74(t), 76, 78; compared with France and Canada, 72, 83, 106-9; comprehensiveness (1970s *vs.* 1990s), 77(t), 78-9, 107-8; Conservation Compliance, 74(t), 78; Conservation Reserve Enhancement Program (CREP), 76, 78, 83, 107; Conservation Reserve Program (CRP), 74(t), 76, 78, 83, 107; cross-compliance, 74(t), 77(t), 78; disconnect between federal and state networks, 33, 71, 88-9, 108, 111-112; educational, 74(t), 77(t), 78; Environmental Quality Incentives Program (EQIP), 74(t), 76, 83, 107; farm bills and agricultural policy, 76; financial incentives, 74(t), 76-8, 79, 83, 88, 108; intrusiveness (1970s *vs.* 1990s), 77(t), 78; land grant universities, 74(t), 78; National Pollutant Discharge Elimination System (NPDES) permits, 73, 81(t); policies re manure spreading, 73, 75; regulatory, 73, 74(t), 75-6, 77(t), 79, 88; in 1970s and 1990s, 72, 73, 74(t), 75-9; Sodbuster, 74(t), 78; Swampbuster, 74(t), 78; USDA's financial incentives, 72

Policy instruments, United States (state): command-and-control regulations, 72, 77(t), 79-80, 81(t), 82, 83, 108; comprehensiveness (1970s *vs.* 1990s), 77(t), 82; cross-compliance, 77(t); disconnect between federal and state networks, 33, 71, 88-9, 108, 111-12; educational, 77(t); financial incentives, 77(t), 82, 83, 106; intrusiveness (1970s *vs.* 1990s), 77(t), 82, 106, 108; livestock regulations, 79-80, 81(t), 82, 87, 106; in 1970s and 1990s, 72, 74(t); Section 319 Program, 87

Policy-making performance, assessment: assessing performance, 9-10; Canada, France, US compared, 71, 72, 83, 90, 94, 96, 103-4, 106-9, 113, 120-1; characteristics of policies used for evaluation, 30-3, 34(t), 35; countries chosen for study, 15-18; criteria for high performance, 8, 46, 113; difficulty in comparing data from different countries, 22-3; economic viability of agriculture, 32-3, 34(t), 35; external success criteria, 11-12; network metaphors for policy formulation, 40-1; network structure, influence on performance, 45-6, 67; new politics of the welfare state, 47-9, 116-18; objective-oriented approach, 21-3; policy formulation theories, 40-1; policy outputs *vs.* policy outcomes, 12; principles for evaluating policy success, 12; problem-based approach, 9-10, 21; problems with policy evaluations, 20; role of bureaucracies and interest groups, 5; solution-oriented approach, 23-9, 35; utilitarian principles, 10-11, 121. *See also* Agro-environmental policy development; *entries beginning with* Policy instruments

Policy-making performance, Canada: actor constellations as cause of poor performance, 90, 96-7, 112, 120-1; clientelism in Ontario, 102, 112-13, 118; impact of internationalization, 104, 120-1; low performance, 94, 96, 103-4, 105, 109, 111-13, 120-1
Policy-making performance, France: actor constellations and corporatist network, 69-70, 109-11, 113, 119-20; move to high agro-environmental performance, 54-5, 63-4, 69-70
Policy-making performance, United States: disconnect between state and federal networks, 33, 71, 88-9, 108, 111-12; federal government performance, 71, 79
Policy networks: actor constellations favourable for policy-making performance, 41-6, 105, 109; adaptation to regional or international pressure, 8; adequate network, overview, 6-8; Canadian federal pressure pluralist networks, 97, 99; characteristics and choice of policy instruments, 7-8; clientelist networks, 44-6, 90, 100-4, 112-13, 118; corporatist networks *(see* Corporatist networks); definition, 4; durable patterns of interaction, 43-4; France, Canada, and US compared, 17, 113; France and corporatist policy network, 17, 55, 67-8, 69-70, 109-11, 113, 119-20; governed interdependence, 8, 45, 67, 70; impact of intergovernmentalism, 49-53, 119; importance of, 8, 25, 43; input-oriented legitimization, 6-7; in internationalized environments, 49-53; issue networks, 44, 45; network structure and balance of power, 44-6; network structure and policy-making performance, 45-6, 67; new politics of the welfare state, 47-9, 116; Ontario's clientelist networks (1990s), 90, 100-3, 104, 112-13, 118; Ontario's pressure pluralist networks (1980s), 100; output-oriented legitimization, 6-7, 105, 121; pressure pluralist networks *(see* Pressure pluralist networks); relation to epistemic communities, 25; state-corporatist networks, 44-5; state-directed networks, 44-6, 50-1, 85-6, 111, 119; with strong state agencies, 8; US federal actors and state-directed networks, 85-6, 111-12; US state actors and pressure pluralist networks, 87-8, 111, 113; US state and federal networks disconnected, 33, 71, 88-9, 108, 111-12; USDA's strength and state-directed networks, 86. *See also* Actor constellations
"Pork Dialogue," 85
Power of joint action, 41-2, 109, 110, 122
Pressure pluralist networks: balance of power of actors, 44-6; Canadian federal pressure pluralist networks, 97, 99; in multilevel governance policy environment, 50(t); Ontario's networks (1980s), 100; structure's influence on policy-making performance, 46; tendency toward Europeanization, 52; and US state actor constellations, 87-8, 111, 113
Programme de maîtrise des pollutions d'origine agricole (PMPOA, France), 60(t), 61, 63, 68-9, 108, 132n41
Putnam, Robert D., 3-4, 10-12, 15, 21, 51

Quebec: policy instruments, 94

Rabe, Barry G., 25-6
Rainelli, Pierre, 58
Random windows of opportunity, 39
Regionalism, 8, 120
Rhodes, R.A.W., 7
Risse-Kappen, Thomas, 51

Sabatier, Paul A., 84
Salmon, Jean, 68
Scharpf, Fritz, 6, 7, 41, 42, 43, 52, 55, 68, 109, 120, 121
Schneider, Ann L., 4, 43
Senate Standing Committee on Agriculture, 98, 99, 106
Skogstad, Grace, 41, 100
Sodbuster (US), 74(t), 78
Soil and Water Environmental Enhancement Program (SWEEP, Ontario), 99
Soil at Risk (Canadian Senate report on soil erosion), 14, 106, 124n35
Soil erosion: Canadian Senate report, 14, 106, 124n35
Spillover windows of opportunity, 39-40, 114
St. Lawrence Action Plan Vision 2000 (Canada), 91, 92(t), 93, 135n8
State actors: in corporatist networks, 46, 67, 69-70, 109-10, 117-18; for governed interdependence, 45, 50, 109; in intergovernmentalism, 50-1; need for strong bureaucracies, 46, 70, 110, 118;

in new politics of welfare state, 48, 116-17; in pluralist policy networks, 46, 87-8, 111, 113; role in governance, 4; in state-directed and state-corporatist networks, 45, 51. *See also entries beginning with* Actor constellations
State-corporatist networks, 44-5
State-directed networks: balance of power of players, 44-6; in intergovernmentalist policy environment, 50-1, 119; structure's influence on policy-making performance, 45; and US federal actors, 85-6, 111-12
Stein, Janice Gross, 10-11
Steinmo, Sven, 22
Streeck, Wolfgang, 52
Sutton, Alan, 79
Swampbuster (US), 74(t), 78

United States: agricultural sector, 16; choice for comparative study, 15-18; disconnected federal and state networks, 33, 71, 88-9, 108, 111-12; governance and policy network, 16-17; and new politics of the welfare state, 48, 117; recognition of pollution problem, 13-15, 38, 106, 115. *See also* Actor constellations, United States; Policy instruments, United States
United States Department of Agriculture (USDA), 72, 86-7

Vogel, David, 23

Walkerton, Ontario: water contamination, 15, 96, 102-3, 113
Water pollution: contamination in Milwaukee, Wisconsin, 14; contamination in Walkerton, Ontario, 15, 96, 102-3, 113; cross-media pollution, 14, 25-6, 31-2, 33; French Water Policy (1992), 59-60; Nitrate Directive in France, 59, 60(t), 66, 68, 107
Weersink, Alfons, 35
Weiss, Linda, 8, 45, 51, 67, 109, 119
Windows of opportunity, 38-40, 114-15

Zimmerman, Janet B., 25-6
Zones d'excédent structurel (ZES, France), 60

Printed and bound in Canada by Friesens

Set in Stone by Brenda and Neil West, BN Typographics West

Copy editor: Robert Lewis

Proofreader: Deborah Kerr

Indexer: Patricia Buchanan